Modern Biology

ITS CONCEPTUAL FOUNDATIONS

Modern Biology

ITS CONCEPTUAL FOUNDATIONS

Edited with Introductions by

ELOF AXEL CARLSON

Associate Professor of Zoology
University of California, Los Angeles

GEORGE BRAZILLER
NEW YORK

IN MEMORY OF

Amanda Marie Carlson

Published simultaneously in Canada by Ambassador Books, Ltd., Toronto

For information, address the publisher:
George Braziller, Inc.
One Park Avenue, New York, N.Y. 10016
Library of Congress Catalog Card Number: 67-12476
First Printing
Printed in the United States of America

Preface

THIS VOLUME ON THE conceptual foundations of modern biology does not attempt to cover the history of biology. An anthology of such a scope would be misleading because what is taught in the university today has little to do with the thinking, the research, and the observations of biologists before the nineteenth century. To understand contemporary biology the layman or uninitiated student requires a grasp of five major trends or concepts. These themes are discussed under the sections on the cell theory, heredity, development, evolution, and molecular biology. All, except the last, have their major roots in the nineteenth century. It could be argued that these concepts date to antiquity, but this would only be by the coincidental similarity of speculative thinking. Biology as an experimental science is new. Isolated and respectable work exists for the seventeenth and eighteenth centuries but these rarer contributions lacked the provocative and enduring influences of the great nineteenth century insights into the nature of life.

Even more important than the nineteenth-century roots of biological thinking today are the rapid and spectacular advances in biological research during the twentieth century. The century began with the excitement caused by the rediscovery of Mendel's experiments. Heredity was no longer a speculative and philosophic pastime of biologists; it now became a field for experiments and the experiments were productive. The cell theory took on new meaning with the recognition that heredity and development could be analyzed at the cellular level. Evolution theory was stimulated by the implications of the theory of the gene.

Perhaps the most significant feature of twentieth century biol-

ogy is the gradual ascendance of biochemistry and biophysics. The merging of physics, chemistry, and biology has led to the new field of molecular biology. It is this new trend, about fifteen years old, which has created an excitement in biology unrivaled since the announcement of Darwin's theory or the rediscovery of Mendelism. Through the approaches of molecular biology, scientists have demonstrated the chemical basis of heredity, the structure of the genetic material, the genetic control of protein synthesis, the molecular structure of membranes and internal cellular organelles, the structure and chemical composition of proteins, the nature of the genetic code, and the regulation of developmental processes in cells.

I have also omitted from this volume the many contributions of nineteenth and twentieth century biologists to organismic biology, general physiology, ecology, anatomy, and other specialized fields. My decision to do so was based on the belief that the five concepts which I have selected present a more direct approach to the understanding of what life is than do the conclusions derived from the more complex fields dealing with whole organisms among the higher plants and animals.

The selection of articles for these five major themes was difficult. There are so many outstanding contributions that any choice would be arbitrary for a book this size. Many of these articles were used in courses taught at Queen's University, Kingston, Ontario, and the University of California at Los Angeles. Some were assigned as supplements for textbooks; others were analyzed for their historical implications in seminars. The five themes presented in this volume are the core of a course in biology for the nonscience major which will be offered at UCLA beginning in 1967.

The illustrations for these articles were redrawn by Roland Carlson, whose assistance is appreciated. Mrs. Florence Okuyama provided thorough, accurate, and rapid clerical assistance in typing and preparing the manuscript.

E. A. C.

Contents

Modern Biology: Its Conceptual Foundations

Introduction

Biology is primarily a descriptive science. To understand life we must observe life. Few of us, however, view life in material terms. We forget that our organs are composed of tissues and that our tissues, in turn, are composed of cells. We forget, or we do not know, that most plants and animals are composed of cells and that their metabolism involves common molecules—amino acids, nucleotides, carbohydrates, and vitamins. We also overlook the fact that all life, microbial as well as human, involves certain universal phenomena. These phenomena include, at the biological level, heredity and development. At the biochemical level they involve metabolism.

In the twentieth century the descriptive preoccupation of biologists gave way to experimental approaches. This trend to experimentation has never ceased. Instead, biologists have learned to think of life as a chemist or a physicist would think of inorganic matter. This represents a philosophy or viewpoint of biologists which, for want of a better synonym, may be called *mechanism*. The mechanistic attitude dissociates certain (not all) human values from experimental procedures. The mechanistic experimentalist looks on organisms as being composed of cells, and the cells, in turn, are thought to be composed of complex aggregates of macromolecules as well as smaller molecules. This analytical attitude does not prevent the biologist from "being human." He can enjoy the music of Vivaldi or Bartók, the novels of Stendhal or Hesse, and the paintings of Rembrandt or Van Gogh. What is *denied* to the experimental biologist is the view that life itself is beyond human understanding, either because it is "too complex" or is governed by nonmaterial "vitalistic" forces, or that life is "off limits" in the Faustian sense that certain realms of knowledge are too dangerous or too sacred for human inquiry. Similarly, the physicist sees matter as composed of atoms and in

the astronomer's scientific eye the stars of the lover or the Zodiac of the astrologer appear as gravitational clusters of gaseous elements undergoing thermonuclear reactions.

The dissociation of nonmechanistic human values from the study of life gives the biologist *limited but important* information about the nature of life. To understand this limited (mechanistic) view of life, we need certain observations, tools, and concepts.

A. THE FUNDAMENTAL UNITS OF BIOLOGY

There are two biological units which should be included in the common vocabulary of every literate being. The most familiar unit is *the cell.* A typical cell looks superficially like a peach cut in half. A central *nucleus* is the dominant feature of the cell, just as a pit is the dominant feature of the sliced peach. The nucleus of the cell contains the hereditary macromolecules of deoxyribonucleic acid (DNA). When cells grow and produce more cells, they do so by a process called *mitosis.* This process assures that each cell which divides produces two cells which have *identical* hereditary compositions in their nuclei.

Modern electron micrographs reveal a variety of structures in the cell (Fig. 1). The gross structure shows a nucleus surrounded by layers of interconnected membranes. Between these layers are found occasional potato-shaped objects—mitochondria—which are also membranous in structure. These mitochondria carry out the oxidations by which the air we breathe becomes transformed with much of the foods we eat into "fixed" energy and a few simple organic compounds. This process results in an efficient chemical system for the storage and utilization of energy at the cellular level. The layered membranes, frequently called the *endoplasmic reticulum,* provide the machinery with which protein is synthesized. Proteins serve three essential cellular functions. *Enzymatic proteins* can bring about all the chemical reactions for breaking complex molecules into smaller units, or conversely, they can synthesize complex molecules from very simple chemical substances. For example, several dozen enzymes are required in a plant cell to construct a molecule of sugar from water and carbon dioxide. *Structural proteins* usually occur as small subunits of membranes and other portions of cell organelles. A third class of proteins, called *regulatory proteins,* are important because they can switch biochemical activities on or off.

The membranes of the endoplasmic reticulum are rich in *ribosomes.* The ribosomes receive *messenger RNA* (ribonucleic acid) from the nucleus. The interplay of messenger RNA and ribosomes, with other components of the cellular system, is essential for protein synthesis.

Inside the cell is the nucleus, whose membrane surrounds the numerous chromosomes characteristic of the species. The DNA of these chromosomes provides the coded sequence from which messenger RNA may be copied.

The electron micrograph view of the cell also indicates that the cell is not really like a firm peach with a pit in its center; rather, the cell is porous like a sponge and both its surface membrane and its nuclear membrane are riddled with holes. The various channels connecting the endoplasmic reticulum make the cell an open system in which the environment may be both inside and outside the cell.

The second fundamental unit in biology is *the gene.* This is the sequence of nucleotides in a DNA molecule which carries a message. That message may be the eventual sequence of amino acids in an enzyme, a structural protein, or a regulatory protein. It is the function of the genes to produce proteins.

The most significant feature of the gene, first discovered in 1921 by H. J. Muller, is called *convariant reproduction.* Unlike any other class of molecule, the gene may lose its normal function by mutation but that alteration does not change its mode of reproduction. The basis for convariant reproduction was unknown until 1953 when the Watson-Crick model of DNA was proposed.

B. THE FIVE MAJOR CONCEPTS OF BIOLOGY

There are over a million different kinds of plants and animals, and whatever we conclude about life must be based on the examination of only a tiny segment of all this life. For this reason *conceptual generalizations* are used by biologists to unify many of their statements about life.

One such conception is the *cell doctrine.* Virtually all organisms visible to the naked eye and almost all organisms seen under the light microscope are composed of cells. All of the cells in an organism arose from one original cell at conception by a process called *mitosis.* Mitosis involves three major events; the first is the replication of hereditary macromolecules; this is followed by the packaging of these replicated molecules into forms capable of

mechanical distribution; and, finally, the distribution of the packaged material in equal quantity. The packages are the *chromosomes,* threads of DNA, which in packaged form resemble spools of thread. When not in the process of mitosis, the DNA threads are loosely spun out in the *nucleus* of the cell. In the unraveled form, the DNA controls the growth and activity of the cell. For this reason the period between cell divisions is not a "resting stage" but an active metabolic phase of the cell's life history.

A second concept involves *heredity.* This phenomenon requires the storage of specificity ("genetic information") and its transmission to progeny. *Genetic information* represents the total heredity which each individual receives at conception. In man, this information is transmitted from cell to cell by mitosis. The adult body contains about 10^{15} (one quadrillion) cells. The genetic information in each of these 10^{15} cells is the same as that found in the original fertilized egg which gave rise to the adult organism. Special cells, sperm and eggs, transmit the information from one generation to the next. Fertilization of an egg by a sperm results in a unique and complete set of genetic information for a new individual. The macromolecule that stores the information is DNA. The DNA is composed of nucleotides and sequences of these nucleotides form *genes.* Genes are analogous to words formed by sequences of letters, the nucleotides representing an alphabet for the formation of genes. The theory of the gene is the major conceptual contribution of the twentieth century to the phenomenon of heredity. The way genes work has been solved in the past twenty-five years. Although the story is not complete, a "dogma" has been formulated and it has largely been demonstrated to be correct. This dogma asserts that DNA (the gene) carries a code; the code from the DNA in the nucleus is transcribed as RNA (another macromolecule, similar to DNA) and carried to the cell cytoplasm. The RNA is decoded in *ribosomes* by a complex process. The *decoded gene* is called a *protein.* The dogma is frequently cited as DNA→RNA→PROTEIN. There are many philosophical features to this dogma. First, the arrows point in one direction (e.g., proteins cannot transmit their information to RNA). There is no experimental evidence that this must be so, but it is factually correct that the direction *is* one way. Second, a code must be assumed for the process to occur, because DNA is composed of nucleotides and proteins are composed of

amino acids. They are analogous to two foreign languages, with the code representing the translation process.

A third concept is *development.* Organisms have definite form —they may be complex in form like people, but at one time in their life history they were less complex. Usually, they started life as a single cell. Two processes are involved in the change from one cell to a recognizable organism. One process is *epigenesis,* the successive change in complexity accompanying cell divisions after fertilization. Biologists do not know what makes cells divide, but such division is necessary for *growth.* They do not know what brings about *cellular differentiation,* yet such differentiation results in muscle, nerve, blood, epithelium, and other tissues. All of these tissues are composed of cells with the same genetic content as the fertilized egg. Biologists know only a little about the chemical basis of cell movements which align different tissue layers properly, thereby forming complex organs as well as the symmetry and organization of the adult body.

The second process of development is *genetic control.* Here several problems may be considered. If all cells in an individual contain the same genetic information, why are there different tissues? Geneticists interested in development thought that genes do not function all the time. But what causes genes to turn on and off in development? This is not known, but recent theories offer some exciting prospects for an answer. Second, what maintains the constancy of chromosome number from parent to progeny? Another way of asking that question would be, how is the heredity of the parent transmitted to the progeny? A process of *meiosis* accomplishes this. The normal number of chromosomes is called *diploid,* implying, correctly, that we are composed of *two* sets of information. A sperm and an egg at fertilization bring together two similar sets of heredity. A special process occurs in the testes and ovaries of maturing adult animals which reduces that diploid number into two equivalent *haploid* sets of chromosomes. All diploid organisms are composed of two haploid sets of chromosomes, one derived from a sperm, the other from an egg.

Although some biochemical work has been done in experimental embryology, very little is known about the molecular basis of development. By the year 2000, the scientific knowledge about development should be as advanced as it is for genetics today.

A fourth concept to be considered is *molecular biology,* a concept which is probably the most refined consequence of the

mechanistic philosophy. Can life be attributed primarily to the organization of macromolecules of nucleic acid and proteins? Are there cellular phenomena which defy a biochemical and biophysical analysis? If not, can the phenotype (appearance) of an organism eventually be defined by a mere listing of its sequence of DNA nucleotides? The directions of molecular biology are manifold. The artificial synthesis of hereditary material has been achieved by A. Kornberg and his colleagues; the *in vitro* synthesis of protein from genetic material has been demonstrated by many investigators but most definitively by the school of H. Hoagland. The genetic code through which this synthesis of proteins is accomplished was deciphered at the laboratories of M. Nirenberg and S. Ochoa. The artificial synthesis of simple viruses is imminent; already the protein of the simplest known virus has been synthesized *in vitro* from its nucleic acid by S. Spiegelman.

In an applied direction, molecular biology gives medicine a new insight into the parasitization of man by viruses. It provides a basis for studying the nature of cancer, resolving an older controversy whether cancer arises from virus infection or from mutation. The molecular basis of several diseases in man has been reported. In one case (phenylketonuric idiocy) a universal test for the disease among newborn infants is available; the test ("the wet diaper test") can detect the chemical defect. Proper treatment, based on the biochemical understanding of the enzyme failure in this disease, now saves these genetically disadvantaged children from mental retardation. Of course, the child does *not* have his defective genes changed by the treatment, so that such procedures raise moral issues about the responsibility of man to his future kindred.

The final concept to be considered is *evolution*. The relation between fact and theory is usually stressed when evolution is discussed. Opponents of evolution claim that "it's only a theory." True, it is a theory, but so is the history of the United States. We may argue whether a *causal* relation exists in American history as vigorously as we do about the phylogenetic relations of life. Unfortunately, this consistency is not usually found. For example, opponents of evolution accept history as "a fact," yet it is really a conceptual scheme for organizing the past existence of human life on earth. We may reject or accept the *theory* that the Spanish-American war was Hearst's war, or that it was a response of an outraged people to atrocities by Spain, or that it was an imperi-

alist design by a few American politicians. Similarly, we can reject or accept the *theory* that evolution resulted from the inheritance of acquired characteristics, or that it represents the filling in of new environments by sudden mutations in old species leading to new species, or that it is a consequence of the natural selection of variations which arise from gene mutation. But if we accept history, per se, as a "reality," so too we must accept the *fact* of evolution.

This larger philosophical view of evolution often overshadows the scientific problems of evolution. The most significant theory of evolution is "the primacy of the gene" as the basis of evolution. How valid is this view? Can "protoplasm" and other vaguely defined substances be conceptually replaced by informational macromolecules as the significant material for evolution? If so, was the first life a gene or at least a DNA molecule? Does the gene alone provide a basis for the origin of variation?

C. FUNDAMENTAL BIOLOGICAL PROCESSES

There are four fundamental biological processes which have already been mentioned—mitosis, meiosis, genetic decoding, and metabolism. *Mitosis* is the process by which one cell appears to divide and form two cells. The entire genetic contents of the cell are replicated, packaged, and transmitted so that each daughter cell receives a complete set of the sequences of nucleotides composing all the gene messages of the parent cell. The process of mitosis involves two aspects. One, the genetic aspect, is the manner of replication and packaging of the chromosomes. The other, the physiological aspect, is the chemistry of the mitotic apparatus which distributes the packages of chromosomes properly. The major mitotic features are illustrated in Figure 2.

In *meiosis* there are two cell divisions which transform a cell with a *diploid* chromosome number into one or more gametes containing a *haploid* number of chromosomes. Fertilization, of course, reconstitutes the *diploid* state and it also brings into being a new individual. The first of these meiotic divisions packages the chromosomes (which constitute the diploid state) and also brings them into paired association. The paired chromosomes separate into two haploid sets. The second division, called an equational division, renders the chromosomes into an unduplicated form (see Fig. 3). Meiosis provides the basis for shuffling the two sets of chromosomes derived from the sperm and egg at

conception. The genetic consequences of this shuffling process enable gene mutations from the mother and the father to be brought into all possible combinations with one another so that they can be "tried out" in nature. Natural selection then can serve as a sieve for the survival of the best combinations. Meiosis, in short, provides the *biological significance of sex*. It is not just more progeny, but progeny showing a range of variations that are brought about by meiosis.

Genetic decoding involves several processes. Messenger RNA is transcribed from the genes beginning at a specific starting point. These messages read only for a short length which contains not more than a few genes. The sequence for a termination point of transcription has not been found. The messenger RNA travels to the cytoplasm (the process by which this happens is not known but that it *does* happen is known). In the endoplasmic reticulum the messenger RNA encounters ribosomes. These ribosomes bring together the nucleotide sequences of the messenger RNA with specific (complementary) sequences of *transfer RNA,* each of which carries an amino acid. There are about twenty different amino acids, and for each kind of amino acid there are approximately three different transfer RNA's which can carry it. The amino acids are aligned by the transfer RNA's as they are paired to the messenger RNA by the ribosome. The amino acids are joined to form a peptide or protein. Figures 4 and 5 show the general scheme for genetic coding. Figure 4 illustrates the protein-synthesizing process and the regulatory mechanism which makes the synthesis stop or start. Figure 5 illustrates the *principle of colinearity*. The gene (DNA) is linear, as is the genetic "map" of the gene derived by special genetic techniques. The protein product of the gene is also linear, and all levels of analysis are in agreement that the specificity must reside in the sequence of subunits composing the DNA or protein.

The fourth fundamental biological process is *metabolism*. It is a peculiar but true conclusion that small simple molecules (an amino acid, a vitamin, a nucleotide, a sugar) are made by large complex molecules (enzymes). Usually *several* different enzymes are required to make any of these simple molecules. The process begins in *plant cells* where the gas, carbon dioxide, and water are combined by a complex process in chloroplasts. The process (photosynthesis) requires energy, which is supplied by *sunlight* and retained by *chlorophyll,* the compound which gives the plant its

green color. The result of photosynthesis is twofold—the gas and water form sugar (glucose), and the energy released by this process is stored in chemical form. There are about thirty different steps involved in this process. Once carbon dioxide is chemically transformed into a few compounds such as glucose, the rest of metabolism involves the further synthesis or degradation of these molecules. The various ions and small organic molecules are brought together by enzymes and larger, more complex forms can be constructed. This is not a haphazard process, however, because of the genetic regulatory mechanisms which govern these reactions. There are several hundred biochemical pathways which have been worked out over the past fifty years. Virtually all of the major constituent molecules of macromolecules have been studied and the steps in their synthesis are known.

D. THE CONCEPT OF THE LIFE CYCLE

In the descriptive phase of biology, a major emphasis is placed on the *life cycle* of the organism. In man, for example, the diploid phase begins at fertilization, goes through rapid embryonic cell division, tissue formation, and organ formation during the early embryonic development in *utero,* and concludes the fetal life with a long gestation which enables the newborn to have a functional nervous system, an adequate musculature, and a physiology which can cope with an extrauterine environment. The newborn infant takes another ten to fifteen years before it can (biologically) produce sex cells or gametes for the next generation of diploid progeny. The adult continues the reproductive phase for some time after the onset of puberty and finally the individual dies.

In other forms of life these various phases (diploid or haploid) occupy different proportions of the life cycle. In microbes, the haploid phase may be predominant and the diploid phase, formed shortly after fertilization, may quickly terminate with meiosis. The significance is nevertheless the same: sexuality provides the basis for the *genetic recombination* of parental variations, and it is meiosis that serves as the mechanism which shuffles the genes.

In some forms of life the processes needed for life are reduced to bare essentials. A noncellular, molecular system which carries out a life cycle is the virus. The most well-studied virus, bacteriophage, is a parasite of the bacterium *Escherichia coli.* As a molecular aggregate the virus has no metabolism, that is, it is *not alive* and produces no progeny. In the host cell, however, the

virus *is alive* because its genes are decoded and the metabolism directed by the enzymes formed from the specificity of its genes produces more viral structural proteins and more viral nucleic acid. The interpretation of this viral parasitism may be designated *infectious heredity* (see Fig. 6).

E. THE IMPLICATIONS OF BIOLOGY FOR HUMAN VALUES

The origin of life was probably the molecular formation of nucleic acid. Once "convariant reproduction" occurred, the process of mutation could become part of the evolution of macromolecules. How the first nucleic acids arose is not known, but many of the components can occur *de novo* in simulated sterile atmospheres which are subjected to electric discharges. Amino acids, sugars, and nitrogenous bases have been found following such experiments. The evolution of the machinery for protein synthesis is not known. But much is known about the polymerization and accretion of the synthesized molecules into visible globules. Biologists may succeed in synthesizing a primitive living system (a pseudovirus, for example) but they are much *closer* now to synthesizing *known* viruses.

With a knowledge of the molecular basis of life, biologists face the dilemma imposed by their curiosity and their consciences. What are the values that guide the use of this knowledge? Should *preventive medicine* be extended to the earliest phase of the life cycle—the gamete itself? In the genetic code are the messages that may give one individual a potential for eminence and another chronic illness. Is the guidance of human evolution desirable, or is it a necessity borne out of our own understanding of man's interference with his own evolution, an interference that began with man's gregarious habits and resulted in his unique control of much of his environment?

The moral values faced by the biologist today are as grave as those faced by the physicists at mid-century with the application, for good and evil, of their mastery of atomic energy. The decisions that result from new knowledge are not usually made by the scientists who introduced that knowledge, nor would such decisions be made more wisely by the scientists. But decisions are made by people and an informed people have more alternatives and more rational opinions with which to reach such decisions.

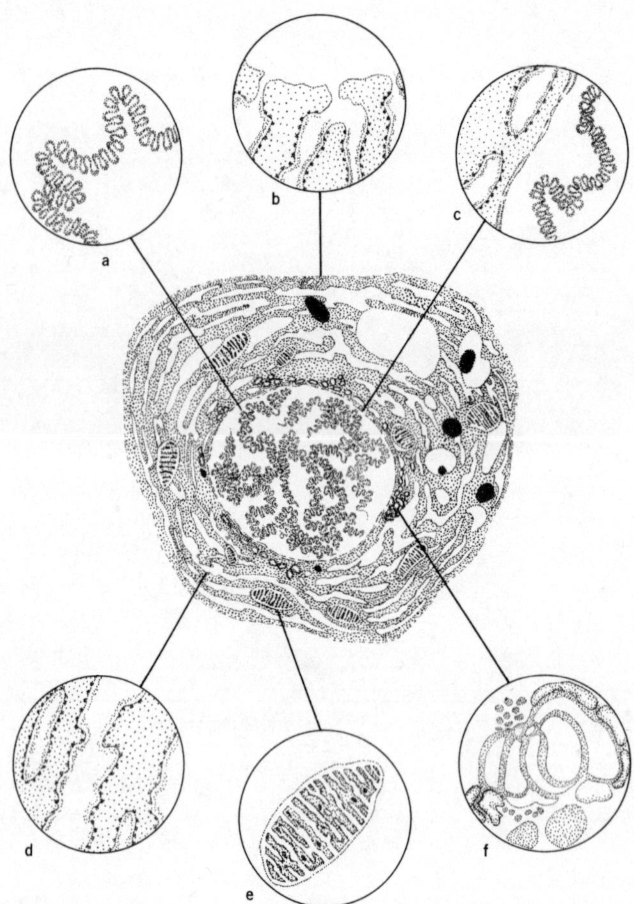

FIG. 1. *The microanatomy of the cell.* The electron microscope has been used widely during the last fifteen years. By it the structure of the cell is revealed as consisting of (a) the DNA of the nuclear chromosomes, which are sometimes seen in the coiled condition; (b) the porous cell membrane, whose openings enter into the endoplasmic reticulum; (c) the nuclear membrane, which, like the cell membrane, has openings into the endoplasmic reticulum; (d) the membranous endoplasmic reticulum, which contains the beadlike ribosomes which carry out protein synthesis; (e) the mitochondria, which oxidize organic compounds and store the energy in molecular form; and (f) the golgi apparatus, a much folded out-pocketing of the endoplasmic reticulum whose secretory functions are still uncertain. The black objects are zymogen granules, which are accretions of enzymes stored in an inactive form.

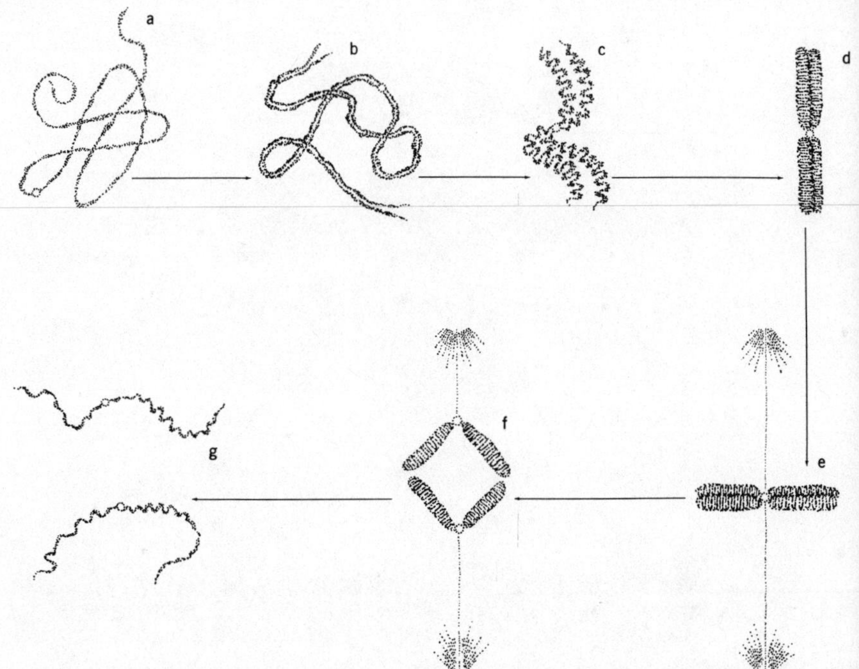

FIG. 2. *The process of mitosis.* The genetic feature of this system involves (a) the metabolism of the DNA thread (chromatid), including the synthesis of nucleotides; (b) the replication of the DNA thread forming two chromatids held by one centromere (round object); (c) the coiling of the two chromatids; (d) the packaged, tightly coiled chromosome, which looks sausage-shaped; (e) the alignment of the chromosome equidistant between the two poles (rayed areas) of the physiological apparatus (spindle) for movement of the chromosomes; (f) the separation of the centromere into two parts, moving in opposite directions along the length of a spindle fiber; and (g) the uncoiling and return to the metabolic condition of the two chromosomes, each functioning in the separate daughter cell which resulted when the parent cell was pinched in two after the chromosomes had moved to opposite poles.

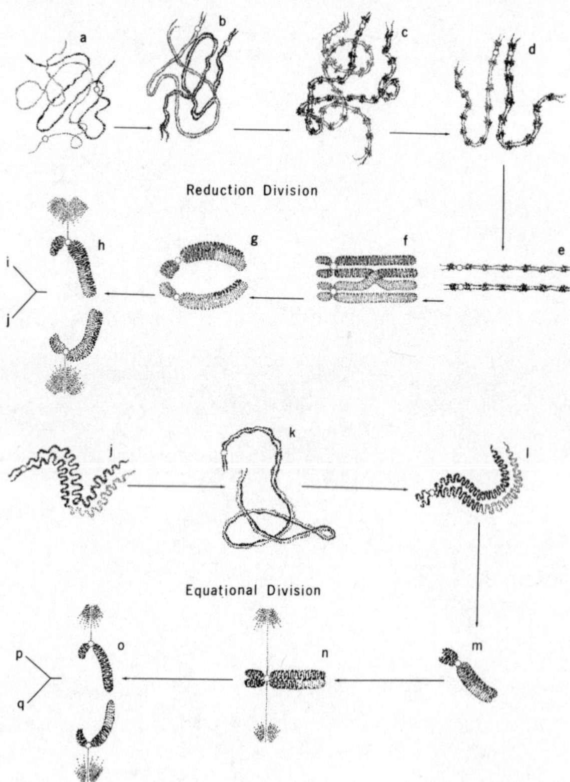

FIG. 3. *The process of meiosis.* The first division is shown with (a) *two* chromosomes that contain the same sequence of genes; one is shown darker than the other. These two chromosomes are called *homologous chromosomes;* (b) the replication of the DNA of each homologue results in the formation of two chromatids per chromosome; (c) the partial coiling of these chromatids yields beadlike chromomeres; (d) the homologues pair by a process called synapsis. The chromomeres match one another in the homologous chromosomes; (e) the completely paired chromosomes; (f) an exchange of chromatids in the pair of homologues; such exchanges result in crossing over. Note the transfer of genes from one homologue to the next in (g), where the continued coiling makes the chromatids more apparent; (h) the movement apart of the homologues resulting in the actual reduction division from the diploid to the haploid state. Two cells, (i) and (j) will form. In each of these cells, an equational division occurs. In (j) the two chromatids uncoil, and (k) the uncoiled chromatids again coil (l) until the compact chromosome (m) is aligned between the spindle poles (n). The separation of the two chromatids in (o) constitutes the equational division into (p) and (q).

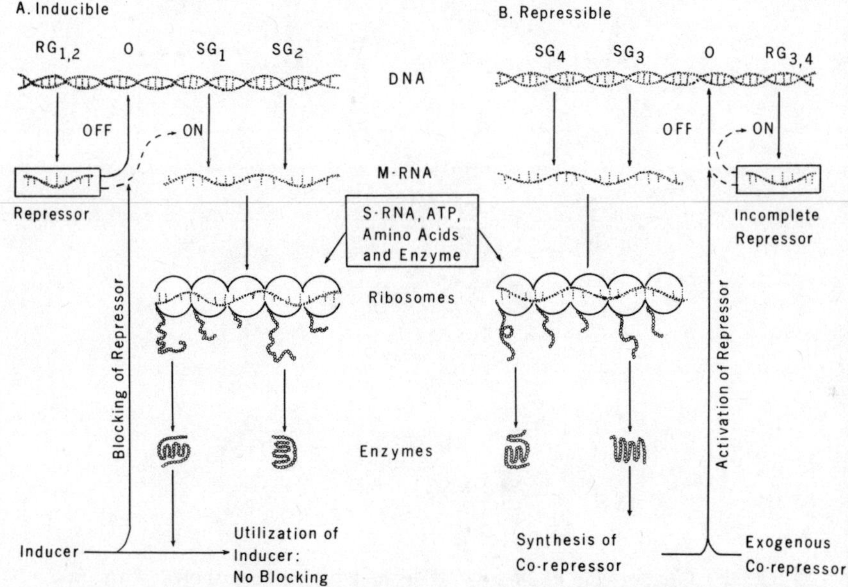

FIG. 4. *The regulation of protein synthesis.* In A, structural genes (SG_1, SG_2) are coordinated by an operator region (o) which is activated by a regulatory gene ($RG_{1,2}$). The regulatory protein (repressor) combines with the operator and prevents genes SG_1 and SG_2 from being transcribed. In the presence of a metabolite (e.g., a sugar on which the enzymes of SG_1 and SG_2 can act), the repressor fails to combine with the operator and the genes SG_1 and SG_2 are transcribed. The messenger RNA for SG_1 and SG_2 is decoded in the ribosomes and the enzymes produced utilize the metabolite. As the metabolite is used up, the repressor is no longer blocked from the operator and the genes are switched off. This system of feedback is called *inducible enzyme synthesis.*

In B, structural genes SG_3 and SG_4 remain on unless the regulatory protein $RG_{3,4}$ is completed by combination with a co-repressor (e.g., an amino acid synthesized by SG_3 and SG_4). The messenger RNA is decoded and the enzymes synthesize the co-repressor which shuts off the operator. Such a feedback system is called *repressible enzyme synthesis.*

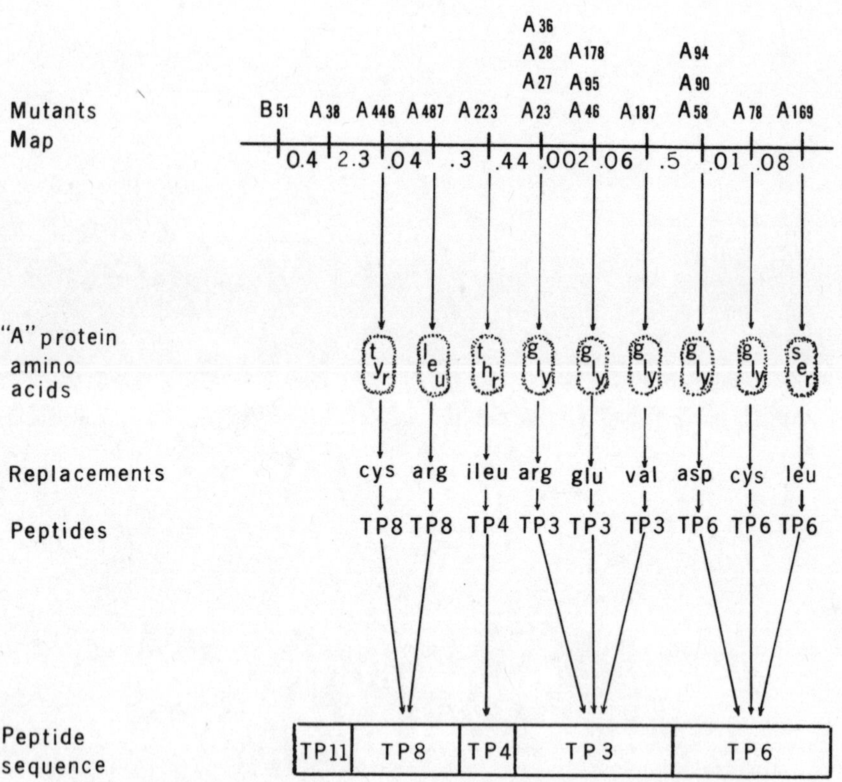

FIG. 5. *The principle of co-linearity.* The protein, tryptophan synthetase, has a sequence of amino acids. Segments of this sequence (peptides) are arranged in the order TP11 TP8 TP4 TP3 TP6. Mutations causing defective tryptophan synthetase frequently replace an original amino acid in the sequence of a peptide with some other kind of amino acid. The location of these substitutions is indicated in the different peptides. When a genetic map is made, the sites of mutation are in the same linear sequence as the sites of amino acid replacement in the peptides of the intact enzyme. Since the DNA is the basis for the mutant map and also for the specificity of the enzyme tryptophan synthetase, it is apparent that a linear molecule DNA generates a linear map and is decoded into a linear sequence of amino acids. The proof that this is the case was accomplished by C. Yanofsky and his colleagues at Stanford University. It was predicted several years earlier by Crick in his 1957 paper "On Protein Synthesis."

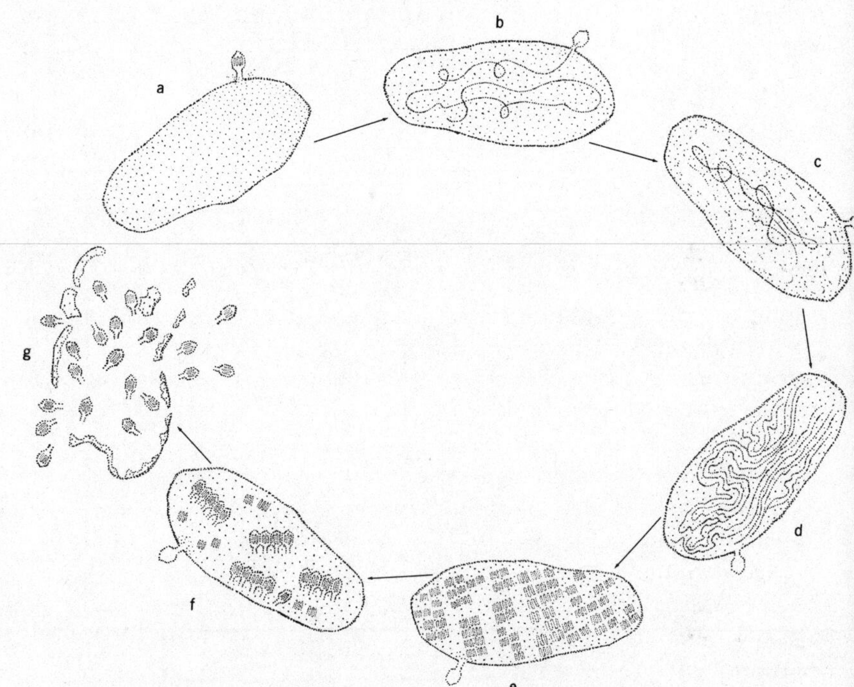

FIG. 6. *The life cycle of the bacteriophage.* The host cell, *Escherichia coli* is several hundred times as large as the T4 bacteriophage. The virus (a) attaches to the host cell, tail first, and (b) injects its DNA into the host; the protein coat of the virus remains outside as a "ghost." In (c) the DNA of the virus is decoded by the host cell and the transcribed messenger RNA enters host cell ribosomes. The enzymes formed parasitize the host and its DNA is degraded into nucleotides. The pool of nucleotides formed by this enzymatic action is converted (d) into a pool of bacteriophage DNA by another virus enzyme which initiates the replication of the virus DNA. The DNA condenses (e) into crystalloid masses around which (f) head and tail fibers are assembled from the protein subunits synthesized by viral structural genes. Another virus-produced enzyme, lysozyme, digests the host cell from within and (g) the bacterium bursts (lysis) releasing one hundred or more progeny viruses.

PART I ○○○

The Cell Theory

THE RECOGNITION OF CELLS required an instrument—the microscope. Anton van Leeuwenhoek, an amateur scientist in Holland, prepared a number of lenses which could magnify up to one hundred and eighty times what the eye could see. In a series of letters to the Royal Society of London (ca. 1674) Leeuwenhoek described his observations—among his discoveries were the "animalcules" which were present in a drop of water and the peculiar microbes extracted from the scrapings of his gums. Somewhat earlier, when Robert Hooke used the compound microscope for viewing small objects, his examination of a piece of cork revealed a porous composition. Because of their cubical appearance, Hooke called the holes he saw *cells.* Later on, it was recognized that the holes observed by Hooke were actually the outer, dead remnants of cells. The *cell* as a unit, however, was not recognized in other forms of life until a century later when R. Dutrachet, M. Schleiden, and T. Schwann claimed that all living things were composed of cells. The verification of this generalization, however, was hampered by the lack of techniques. The staining of tissues was not developed until the 1850's by Gerlach, while the sectioning of tissues for slides was a development of German anatomists in the period 1850–1875. The rapid development of modern microscopic techniques was also stimulated by the *cell theory.* According to R. Virchow, the observation that plants and animals are composed of cells had an important corollary: all cells must *arise* from other cells just as all life must arise from preceding life.

The cell theory was essentially correct. All higher plants and animals are composed of cells. The viruses and some small microbes have defied a cellular definition, and some organisms such

as the true slime molds lack partitioned cells. In these cases, there may be some doubt about their cellular status. Despite these trivial exceptions, the cell theory is universally accepted. It has provided the basis for studies of reproduction, of cell division, of embryonic development, and it has permitted the eventual acceptance of the mechanics of heredity.

In the twentieth century the light microscope provided a good deal of knowledge about the processes of cell division, reproduction, and the role of the chromosomes in heredity. But it is the widespread use of the electron microscope that has transformed the interpretation of the cell. For while the light microscope cannot enlarge an object more than 2000-fold, the electron microscope magnifies the object 100,000 or more times its original size. This enormous magnification, accompanied by advances in ultrathin sectioning and special staining techniques, has revealed a detailed structure of the internal and external components of the cell. Most noticeable is the porous network of membranes which extends from the cell nucleus to the outer covering surface.

Also advancing the knowledge of cell structure and function are the techniques which cell physiologists have developed. Cells can be disrupted into fragments which can be separated according to their weight and chemical composition. This process of *cell fractionation* permits the biochemist to test the function of the various purified particles that are obained. Once these funcions are known experimental variations can be tried out. Many of the advances in biochemistry and molecular biology have been accomplished through such cell fractionation experiments. The synthesis of proteins in test tubes and the isolation of the enzymes involved in photosynthesis or in the oxidation of sugar to produce energy are examples of such fractionation studies.

Robert Hooke

Of the Schematisme or Texture of Cork, and of the Cells and Pores of some other such Frothy Bodies

Robert Hooke was one of those versatile geniuses whose mind seemed at ease in all fields of knowledge. Hooke is perhaps best known as a physicist, but his construction of a compound microscope led to a popularization of microscopy. The *micrographia* carried details of small insects and other objects, but Hooke never obtained the detailed observations of the microscopic world which were achieved by his contemporary Leeuwenhoek in Holland.

Hooke's observation of cork led him to conceive of this substance as a spongy mass of air sacs or "*cells.*" Some two hundred years would elapse, however, before living cells were recognized to be material bodies. Hooke's cells were actually the cell walls of thick cellulose which remained intact after the live cellular material had died.

I took a good clear piece of Cork, and with a Pen-knife sharpened as keen as a Razor, I cut a piece of it off, and thereby left the surface of it exceeding smooth, then examining it very diligently with a Microscope, me thought I could perceive it to appear a little porous; but I could not so plainly distinguish them, as to be sure that they were pores, much less what Figure they were of: But judging from the lightness and yielding quality of the Cork, that certainly the texture could not be so curious, but that possibly, if I could use some further diligence, I might find it to be discernible with a Microscope, I with the same sharp Pen-knife, cut off from the former smooth surface an exceeding thin piece of it, and placing it on a black object Plate, because it

---

Micrographia (London: Martin and Allestry, 1665).

was itself a white body, and casting the light on it with a deep plano-convex Glass, I could exceeding plainly perceive it to be all perforated and porous, much like a Honey-comb, but that the pores of it were not regular; yet it was not unlike a Honey-comb in these particulars.

First, in that it had a very little solid substance, in comparison of the empty cavity that was contained between, as does more manifestly appear by the Figure A and B of the XI Scheme, for the Interstitia, or walls (as I may so call them) or partitions of those pores were near as thin in proportion to their pores, as those thin films of Wax in a Honey-comb (which enclose and constitute the sexangular cells) are to theirs.

Next, in that these pores, or cells, were not very deep, but consisted of a great many little Boxes, separated out of one continued long pore, by certain Diaphragms, as is visible by the Figure B, which represents a sight of those pores split the long-ways.

I no sooner discern'd these (which were indeed the first microscopical pores I ever saw, and perhaps, that were ever seen, for I had not met with any Writer or Person, that had made any mention of them before this) but me thought I had with the discovery of them, presently hinted to me the true and intelligible reason of all the Phaenomena of Cork; As,

First, if I enquir'd why it was so exceeding light a body? my Microscope could presently inform me that here was the same reason evident that there is found for the lightness of froth, an empty Honey-comb, Wool, a Spunge, a Pumice-stone, or the like; namely, a very small quantity of a solid body, extended into exceeding large dimensions.

1665

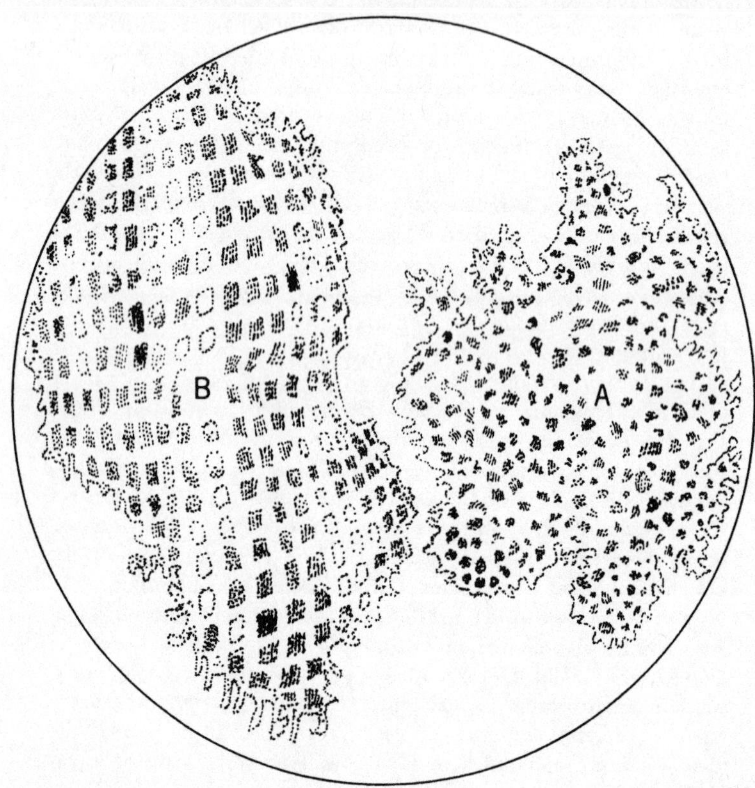

FIG. 1. Hooke's illustration of a piece of cork. A and B represent different sections of sliced cork.

Rudolf Virchow

The Cell Doctrine

Early in the nineteenth century, the botanist M. J. Schleiden and the zoologist Theodore Schwann proposed a structural theory of plant and animal anatomy. They observed that all tissues of living organisms were composed of cells. Like Hooke, they also assumed that the cell was primarily composed of its wall. Only in the mid-nineteenth century was protoplasm recognized as the essential component of the cell. The proof that the cell was a universal structure for living things required better microscopes as well as the technology for preparing tissues for microscopes. One contribution, stain technology, was developed by Gerlach, who discovered how much more detail could be observed when tissue slices were soaked in dyes such as carmine. In the 1850's there was much interest in the cellular composition of tissues. Rudolf Virchow rejected the view that cells emerge *de novo* from rudiments in the cell. He argued that all cells come from preexisting cells. It was this "cell doctrine" which replaced the earlier cell theory of Schleiden and Schwann. The universality of cellular structure in virtually all organisms observable under the microscope led to an intense interest in the detailed structure of the individual cell. Virchow alludes to Brown's observations of the nucleus of the cell. Microscopists of the last half of the nineteenth century, accepting Virchow's cell doctrine, exerted their best efforts to achieve a detailed description of the cell. In so doing they stimulated the fields of embryology and heredity. It is too often overlooked that the cell doctrine came into being at about the same time (1858) as Darwin's *Origin of the Species* (1859). The cell theory provided the basis for experimental cell research, experimental embryology, and, eventually, the experimental evidence for the physical basis of heredity. Note that Virchow refers to Schleiden and Schwann's cell theory as

---

Translated by C. M. Stern. Reprinted from August P. Suner (ed.), *Classics of Biology* (New York: Philosophical Library, 1955), by permission of the publisher.

the "free formation" of cells, a concept which he rejects. In its place he advocates the "cell doctrine"—all cells are derived from preexisting cells. Today the term "cell theory" is used synonymously with the term "cell doctrine."

At the beginning of the present century, Bichat established the principles of General Anatomy, and new horizons were opened to medical science. But the progress which histology owes to Schwann has not been greatly developed nor sufficiently applied to pathology. It is a matter for surprise to see how little has been achieved in the period since that time.

Each day brings forth fresh discoveries but it also opens up fresh matters for uncertainty. Is anything positive in histology, we have to ask ourselves. What are the parts of the body whence commence the vital actions? Which are the active elements, and which the passive? These are queries which have given rise to great difficulties, dominating the field of physiology and pathology, and which I have solved by showing that "the cell constitutes the true organic unit," that it is the ultimate irreducible form of every living element, and that from it emanate all the activities of life both in health and in sickness. This manner of holding life to be a special process may be thrown in my face, and many people may incline to accuse me of some type of biological mysticism, impelling me to set life apart from the great agglomeration of natural occurrences and discount the sovereign laws of chemistry and physics. The object of this course is to show that it is not possible to possess more mechanical ideas than those I myself profess, provided that one seeks to explain what happens in the elementary forms of organism. There is no doubt that the molecular changes occurring inside the cells are referred to some part or other composing them; the final result is, however, due to the cell from which the vital action started and the living element is not active except in so far as it presents us with a complete whole enjoying its own separate existence.

It has many times been objected that there is no general agreement as to what must be understood by the term "cell." These difficulties date from the work of Schwann himself, who, founding his system on that of Schleiden, interpreted his observations as a botanist in such a way that theories of plant physiology were applied to animal physiology without taking into account that even within the general structure there were plant cells presenting peculiar characters not safely to be identified with the characters exhibited by animal cells. Robert Brown's discoveries contributed greatly to the unification of concepts: I refer to his detection of the nucleus in animal and plant cells. An unimportant role was, however, attributed to the nucleus in the preservation of the cell, whilst in contrast its effect in the development of cellular elements was greatly exaggerated.

On the other hand the nucleus, in elements of development, frequently includes an important formation: the nucleolus. According to Schleiden, whose opinion was later adopted by Schwann, the development of the cell takes place in the following manner. The nucleolus becomes the first vestige of tissue, coming to life in the innermost recesses of the formative liquid, blasteme or cytoblasteme, and rapidly acquiring definite volume; small granules separate out from the blasteme and settle around it whilst a membrane is formed to cover these elements. Once the nucleus is completed, a new mass starts to surround it. Some time after, a second membrane starts to form at a definite point on the surface of the nucleus, the well-known "watchglass" form. From this is developed the protoplasm and then the cell membrane.

This theory, admitting the development of the cell at the expense of the free blasteme and requiring the formation of the cell to precede that of the nucleus or cytoblast, is generally known under the name of the "Cell Theory." It would be more precise to call it the theory of the free formation of cells. Nowadays it has been almost entirely abandoned.

In spite of most careful investigations, starting with study of tissues under the microscope, there has never been observed any part which, being ready to grow, or multiply physiologically or pathologically, has not contained nucleate elements as the starting-point of the growth process. In all cases, the first important modifications have their root in the nucleus and this can be confirmed by examining the appearance of the nucleus when an element is about to become transformed. Nothing fresh is created, neither in the shape of complete organisms nor individual elements.

Similarly, the digestive juices do not form the tapeworm, nor do infusoria, algae or cryptogams arise from organic plant or animal waste; in the same way in histology, we deny the possibility of forming a cell at the expense of a non-cellular substance.

The existence of a cell presupposes the prior existence of some other cell—*omnis cellula e cellula* (every cell must come from some other cell)—just the same as a plant cannot occur except it be derived from some other plant, or an animal from some other animal. Throughout the range of living beings, plants, animals, or the constituent parts of one or the other, there rules the eternal law of continuous development. Development cannot cease to be continuous, because no particular generation can start a fresh series of developments. We must reduce all tissues to a single simple element, the cell. The whole of an individual made up of distinct tissues is the result of cellular proliferation. There is only one way to form cells, that is by fissiparity; one element is divided after the other. Each new generation proceeds from some preceding generation.

It is not possible to look on an elementary granule, globule, or fibre as the starting point of histological development; indeed there is no reason to suppose that living elements arise from unorganized parts, any more than it can be assumed that certain substances, liquids, plastic elements, matter or blastema give rise to the formation of cells. The formative materials are to be found within the cell, they are neither extracellular nor precellular.

Such are the observations which seem to me to constitute the starting-point for any theory of biology. A single elementary form makes up the living body and remains ever constant. This brings us to a consideration of the higher formations, the plant and the animal, as the sum of the greater or less number of like or dissimilar cells each possessing all the characteristics necessary to life. It is not, however, in a point of superior organization, in man's brain for instance, that the character of vital unity is to be found, but rather in the regular and constant co-ordination of isolated elements. Thus, the individual is the result of a type of social organization, an agglomeration of individual existences dependent one on the other. But this dependence is of such nature that each element has its own special type of activity and even when other parts produce some motivation on it, some impulse or excitation whatever they may happen to be, the function does not emanate the less from the element itself, because of that.

1858

Edmund B. Wilson

The Cell in Development and Inheritance

E. B. Wilson was a student of W. K. Brooks, who also had T. H. Morgan as a student at Johns Hopkins University. These two students were eventually to come together as members of the same department (Zoology, at Columbia University) and collaborate, as teachers, in the development of the chromosome theory of inheritance. Wilson enjoyed miscroscopy. He saw in the cell theory a basis for an experimental analysis of development and a basis for understanding heredity. Wilson's book *The Cell in Development and Inheritance* appeared in 1896. It was a provocative book which gave the students at Columbia a strong incentive to study the problems of heredity. One of Wilson's students, Sutton, related the movements of chromosomes to Mendelism and Wilson popularized this view in his lectures, articles, and classes.

Wilson was chairman of the Zoology Department when he brought Morgan to Columbia. He encouraged Morgan's researches with the fruit fly, *Drosophila,* although Morgan was by training and temperament an embryologist. Wilson was the theoretician who set the viewpoint for the "Columbia School" of biology. That viewpoint required the joining together of breeding analysis (Mendelism), cytology, and physiology for the analysis of heredity and development. It is a remarkable parallel that these very three requirements were used in the 1960's in the molecular analysis of development by Jacob and Monod!

All of the facts reviewed . . . converge, I think, to the conclusion drawn by Claude Bernard, that the nucleus is the formative centre of the cell in a chemical sense, and through this is the especial seat of the formative energy in a morphological sense. That the nucleus has such a significance in synthetic metabolism is proved

---

Reprinted from *The Cell in Development and Inheritance* (New York: The Macmillan Company, 1896), pp. 261–263, 326–330.

by the fact that digestion and absorption of food, growth, and secretion cease with its removal from the cytoplasm, while destructive metabolism may long continue as manifested by the phenomena of irritability and contractility. It is indicated by the position and movements of the nucleus in relation to the food-supply and to the formation of specific cytoplasmic products. It harmonizes with the fact, now universally admitted, that active exchanges of material go on between nucleus and cytoplasm. The periodic changes of staining-capacity undergone by the chromatin during the cycle of cell-life, taken in connection with the researches of physiological chemists on the chemical composition and staining-reactions of the nuclein-series, indicate that the substance known as nucleic acid plays a leading part in the constructive process. During the vegetative phase of the cell this substance appears to enter into combination with proteid or albuminous substance to form a nuclein. During its mitotic or reproductive phase the albumin is split off, leaving the substance of the chromosomes as nearly pure nucleic acid. When this is correlated with the fact that the sperm-nucleus, which brings with it the paternal heritage, likewise consists of nearly pure nucleic acid, the possibility is opened that this substance may be in a chemical sense not only the formative centre of the nucleus but also a primary factor in the constructive processes of the cytoplasm.

The rôle of the nucleus in constructive metabolism is intimately related with its rôle in morphological synthesis and thus in inheritance; for the recurrence of similar morphological characters must in the last analysis be due to the recurrence of corresponding forms of metabolic action of which they are the outward expression. That the nucleus is in fact a primary factor in morphological as well as chemical synthesis is demonstrated by experiments on unicellular plants and animals, which prove that the power of regenerating lost parts disappears with its removal, though the enucleated fragment may continue to live and move for a considerable period.

This fact establishes the presumption that the nucleus is, if not the actual seat of the formative energy, at least the controlling factor in that energy, and hence the controlling factor in inheritance. This presumption becomes a practical certainty when we turn to the facts of maturation, fertilization, and cell-division. All of these converge to the conclusion that the chromatin is the

most essential element in development. In maturation the germ-nuclei are by an elaborate process prepared for the subsequent union of equivalent chromatic elements from the two sexes. By fertilization these elements are brought together and by mitotic division distributed with exact equality to the embryonic cells. The result proves that the spermatozoön is as potent in inheritance as the ovum, though the former contributes an amount of cytoplasm which is but an infinitesimal fraction of that supplied by the ovum. The centrosome, finally, is excluded from the process of inheritance, since it may be derived from one sex only.

DEVELOPMENT, INHERITANCE, AND METABOLISM

In bringing the foregoing discussion into more direct relation with the general theory of cell-action we may recall that the cell-nucleus appears to us in two apparently different roles. On the one hand, it is a primary factor in morphological synthesis and hence in inheritance, on the other hand an organ of metabolism especially concerned with the constructive process. These two functions we may with Claude Bernard regard as but different phases of one process. The building of a definite cell-product, such as a muscle-fibre, a nerve-process, a cilium, a pigment-granule, a zymogen-granule, is in the last analysis the result of a specific form of metabolic activity, as we may conclude from the fact that such products have not only a definite physical and morphological character, but also a definite chemical character. In its physiological aspect, therefore, inheritance is the recurrence, in successive generations, of like forms of metabolism; and this is effected through the transmission from generation to generation of a specific substance or idioplasm which we have seen reason to identify with chromatin. This remains true however we may conceive the morphological nature of the idioplasm—whether as a microcosm of invisible germs or pangens, as conceived by De Vries, Weismann, and Hertwig, as a storehouse of specific ferments as Driesch suggests, or as complex molecular substance grouped in micellae as in Nägeli's hypothesis. It is true, as Verworn insists, that the cytoplasm is essential to inheritance; for without a specifically organized cytoplasm the nucleus is unable to set up specific forms of synthesis. This objection, which has already been considered from different points of view, both by De Vries and Driesch, disappears as soon as we regard the egg-cytoplasm as itself a product of nuclear activity;

and it is just here that the general role of the nucleus in metabolism is of such vital importance to the theory of inheritance. If the nucleus be the formative centre of the cell, if nutritive substances be elaborated by or under the influence of the nucleus while they are built into the living fabric, then the specific character of the cytoplasm is determined by that of the nucleus, and the contradiction vanishes. In accepting this view we admit that the cytoplasm of the egg is, in a measure, the substratum of inheritance, but it is so only by virtue of its relation to the nucleus, which is, so to speak, the ultimate court of appeal. The nucleus cannot operate without a cytoplasmic field in which its peculiar powers may come into play; but this field is created and moulded by itself. Both are necessary to development; the nucleus alone suffices for the inheritance of specific possibilities of development.

PREFORMATION AND EPIGENESIS. THE UNKNOWN FACTOR IN DEVELOPMENT

We have now arrived at the furthest outposts of cell-research; and here we find ourselves confronted with the same unsolved problems before which the investigators of evolution have made a halt. For we must now inquire what is the guiding principle of embryological development that correlates its complex phenomena and directs them to a definite end. However we conceive the special mechanism of development, we cannot escape the conclusion that the power behind it is involved in the structure of the germ-plasm inherited from foregoing generations. What is the nature of this structure and how has it been acquired? To the first of these questions we have as yet no certain answer. The second question is merely the general problem of evolution stated from the standpoint of the cell-theory. The first question raises once more the old puzzle of preformation or epigenesis. The pangen hypothesis of De Vries and Weismann recognizes the fact that development is epigenetic in its external features; but like Darwin's hypothesis of pangenesis, it is at bottom a theory of preformation, and Weismann expresses the conviction that an epigenetic development is an impossibility. He thus explicitly adopts the view, long since suggested by Huxley, that "the process which in its superficial aspect is epigenesis appears in essence to be evolution in the modified sense adopted in Bonnet's later writings; and development is merely the expansion of a potential organism or 'original preformation' according to fixed laws."

Hertwig ('92), while accepting the pangen hypothesis, endeavours to take a middle ground between preformation and epigenesis, by assuming that the pangens (idioblasts) represent only cell-characters, the traits of multicellular body arising epigenetically by permutations and combinations of these characters. This conception certainly tends to simplify our ideas of development in its outward features, but it does not explain why cells of different characters should be combined in a definite manner, and hence does not reach the ultimate problem of inheritance.

What lies beyond our reach at present, as Driesch has very ably urged, is to explain the orderly rhythm of development—the co-ordinating power that guides development to its predestined end. We are logically compelled to refer this power to the inherent organization of the germ, but we neither know nor can we even conceive what this organization is. The theory of Roux and Weismann demands for the orderly distribution of the elements of the germ-plasm a prearranged system of forces of absolutely inconceivable complexity. Hertwig's and De Vries's theory, though apparently simpler, makes no less a demand; for how are we to conceive the power which guides the countless hosts of migrating pangens throughout all the long and complex events of development? The same difficulty confronts us under any theory we can frame. If with Herbert Spencer we assume the germ-plasm to be an aggregation of like units, molecular or supra-molecular, endowed with predetermined polarities which lead to their grouping in specific forms, we but throw the problem one stage further back, and, as Weismann himself has pointed out, substitute for one difficulty another of exactly the same kind.

The truth is that an explanation of development is at present beyond our reach. The controversy between preformation and epigenesis has now arrived at a stage where it has little meaning apart from the general problem of physical causality. What we know is that a specific kind of living substance, derived from the parent, tends to run through a specific cycle of changes during which it transforms itself into a body like that of which it formed a part; and we are able to study with greater or less precision the mechanism by which that transformation is effected and the conditions under which it takes place. But despite all our theories we no more know how the properties of the idioplasm involve the properties of the adult body than we know how the properties of hydrogen and oxygen involve those of water. So long as

the chemist and physicist are unable to solve so simple a problem of physical causality as this, the embryologist may well be content to reserve his judgment on a problem a hundredfold more complex.

The second question, regarding the historical origin of the idioplasm, brings us to the side of the evolutionists. The idioplasm of every species has been derived, as we must believe, by the modification of a pre-existing idioplasm through variation, and the survival of the fittest. Whether these variations first arise in the idioplasm of the germ-cells, as Weismann maintains, or whether they may arise in the body-cells and then be reflected back upon the idioplasm, is a question on which, as far as I can see, the study of the cell has not thus far thrown a ray of light. Whatever position we take on this question, the same difficulty is encountered; namely, the origin of that co-ordinated fitness, that power of active adjustment between internal and external relations, which, as so many eminent biological thinkers have insisted, overshadows every manifestation of life. The nature and origin of this power is the fundamental problem of biology. When, after removing the lens of the eye in the larval salamander, we see it restored in perfect and typical form by regeneration from the posterior layer of the iris, we behold an adaptive response to changed conditions of which the organism can have had no antecedent experience either ontogenetic or phylogenetic, and one of so marvellous a character that we are made to realize, as by a flash of light, how far we still are from a solution of this problem. It may be true, as Schwann himself urged, that the adaptive power of living beings differs in degree only, not in kind, from that of unorganized bodies. It is true that we may trace in organic nature long and finely graduated series leading upward from the lower to the higher forms, and we must believe that the wonderful adaptive manifestations of the more complex forms have been derived from simpler conditions through the progressive operation of natural causes. But when all these admissions are made, and when the conserving action of natural selection is in the fullest degree recognized, we cannot close our eyes to two facts: first, that we are utterly ignorant of the manner in which the idioplasm of the germ-cell can so respond to the play of physical forces upon it as to call forth an adaptive variation; and second, that the study of the cell has on the whole seemed

to widen rather than to narrow the enormous gap that separates even the lowest forms of life from the inorganic world.

I am well aware that to many such a conclusion may appear reactionary or even to involve a renunciation of what has been regarded as the ultimate aim of biology. In reply to such a criticism I can only express my conviction that the magnitude of the problem of development, whether ontogenetic or phylogenetic, has been underestimated; and that the progress of science is retarded rather than advanced by a premature attack upon its ultimate problems. Yet the splendid achievements of cell-research in the past twenty years stand as the promise of its possibilities for the future, and we need set no limit to its advance. To Schleiden and Schwann the present standpoint of the cell-theory might well have seemed unattainable. We cannot foretell its future triumphs, nor can we repress the hope that step by step the way may yet be opened to an understanding of inheritance and development.

1896

M. Zalokar

Nuclear Origin of Ribonucleic Acid

Many of the observations on cell physiology have required elaborate biochemical and biophysical tools. In this paper, Zalokar uses a very simple system to provide a profound result. Radioactive uridine is a component of ribonucleic acid (RNA); when fed to filamentous molds, the uridine is incorporated by the cell. Radioactivity can be detected on a photographic emulsion, but if the RNA is dispersed throughout the cell, the entire cell would appear to be labeled. By centrifugation of the cells, Zalokar was able to separate the various cellular components into layers within the live cells. A comparison of such centrifuged hyphae, prepared at different times following the incorporation of labeled uridine, clearly showed that RNA is first synthesized in the nucleus and then passes out to the cytoplasm, primarily in the microsomes or ribosomes. Ribosomes are part of the endoplasmic reticulum and they serve as factories for the synthesis of proteins. It is likely that much of the RNA synthesized by the nucleus in this experiment is messenger RNA. The messenger RNA in the ribosomes is decoded to yield specific proteins, usually enzymes.

Living hyphae of *Neurospora crassa* were centrifuged and their contents stratified in distinct layers. Starting at the centrifugal end, the layers were as follows (Fig. 1): glycogen, ergastoplasm (microsomes), mitochondria, nuclei, 'supernatant' cytoplasm, vacuoles and fat. Each fraction could be identified by cytochemical reactions. It was found that most of the cytoplasmic ribonucleic acid resided in ergastoplasm, some in mitochondria, and none was detectable in the 'supernatant'. Nuclei were relatively poor in ribonucleic acid.

The fact that nuclei became clearly separated from the cytoplasmic ribonucleic acid enabled us to localize the site of forma-

Reprinted from *Nature 183*:1330, 1959, by permission of the publisher and the author.

tion of ribonucleic acid. Mycelium, in its active growth phase, was fed tritiated uridine (uridine-5, 6-^{3}H, 640 mc. per m.mole, 100 μgm./ml.) for given times, centrifuged and fixed. The mycelium was then washed in cold 5 per cent trichloracetic acid so that only nucleosides which were incorporated into ribonucleic acid remained. The preparation was mounted on a slide and covered with photographic emulsion.[1] After exposure and development, the autoradiographs showed tracks of β-decay at the sites of uptake of the precursor.

At short feeding times, from 1 to 4 min., the label appeared only in the nuclear fraction (Fig. 2). After 4 min., more and more label was found in cytoplasmic ribonucleic acid. In 1 hr., ergastoplasm was labelled much more heavily than the nuclei (Fig. 3), and some activity was found also in mitochondria, while the 'supernatant' remained virtually inactive. In brief, the distribution of the label became roughly proportional to the relative content of ribonucleic acid in the different fractions. When tritiated uridine was fed for 1 min. only and then washed out and replaced with an excess of non-labelled uridine, the label again appeared first in nuclei, to appear in the ergastoplasm only after several minutes. The labelling of ergastoplasm increased with time at the expense of nuclear label.

These results demonstrate that all the cellular ribonucleic acid is formed in nuclei and that it migrates into the cytoplasm later. Previous work of several investigators indicated that ribonucleic acid originated in nuclei[2] and the present experiments substantiate this hypothesis. These findings suggest that ribonucleic acid is a direct product of gene action. Ribonucleic acid is formed in nuclei, the seat of chromosomes and genes; it migrates into the cytoplasm; and it is required for the synthesis of proteins. A detailed report of this investigation will be published elsewhere.

This work was supported by a grant from the National Institutes of Health, U.S. Public Health Service.

1959

NOTES

[1] Ficq, A., Arch. Biol. *66* 509 (1955).

[2] Brachet, J., *Chemical Cytology* (New York: Academic Press, 1957).

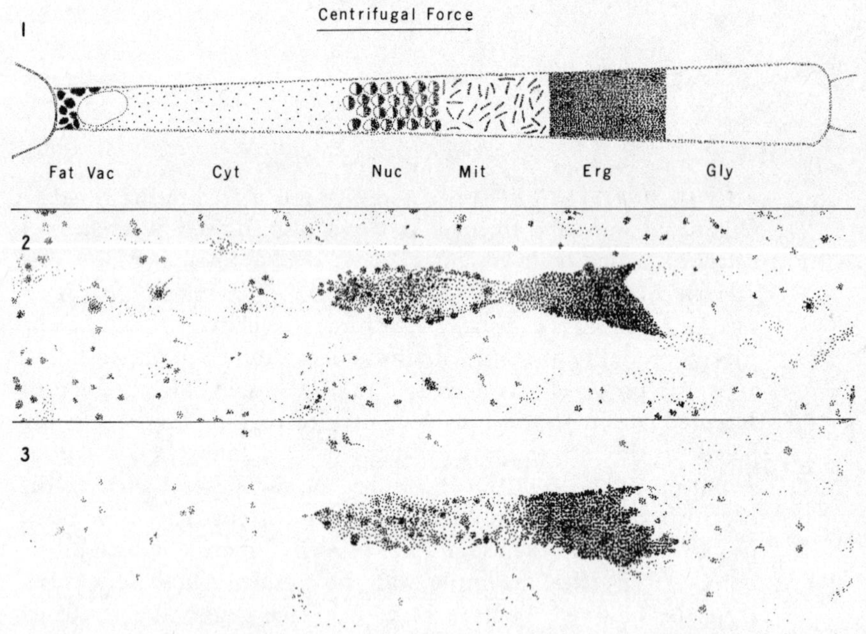

FIG. 1. Schematic presentation of a centrifuged hypha of Neurospora. FAT, fat; VAC, vacuole; CYT, 'supernatant' cytoplasm; NUC, nuclei; MIT, mitochondria; ERG, ergastoplasm (microsomes); GLY, glycogen. Centrifugal direction is to the right in all figures.

FIG. 2. Autoradiograph of a centrifuged hypha fed tritiated uridine for 1 min. prior to centrifugation, stained with haemalum. β-tracks in the nuclear layer; deeply stained layer, free of tracks, is ergastoplasm; glycogen and 'supernatant' not stained.

FIG. 3. Autoradiograph of a centrifuged hypha fed tritiated uridine for 1 min. and unlabelled uridine for 1 hr., prior to centrifugation. β-tracks densest in ergastoplasm, less dense in nuclei and mitochondria; no significant increase of tracks over background in glycogen and 'supernatant.'

PART II ooo

Heredity

IN THE NINETEENTH CENTURY heredity was a frustrating mystery. Organisms inherited likeness, but they also showed differences or *variation* from their parents. These two aspects, heredity and variation, were believed to be separate phenomena. Variation was believed to be a new condition somehow arising in the parent. Heredity, by contrast, was believed to be a more profound ancestral property characteristic of the species. Most biologists accepted an environmental role in directing the variations imposed on the progeny, a view derived from Lamarck's theory of evolution in the eighteenth century. According to Lamarck, the organism responds to its environment and strengthens or weakens its various organ systems, and the progeny inherit these modifications. His classical example was the giraffe, whose long neck arose, he believed, because each successive generation stretched its neck further in a struggle to eat leaves from trees. This theory of the inheritance of acquired characteristics was not completely satisfying to Darwin, who was baffled at the range of variation within a population. It was rejected by A. Weismann, who found no relation between trauma and inheritance (e.g., centuries of human circumcision had *not* resulted in the birth of males lacking a prepuce). Instead, Weismann argued that the hereditary substances are protected from environmental influences early in the embryonic development and that the cells which will constitute the next generation are set aside as a *germ plasm.* The rest of the individual constitutes the *soma;* somatic cells do not participate in heredity. Weismann's theory of the continuity of the germ plasm was supplemented by another remarkable concept. How was it, Weismann asked, that the number of chromo-

somes remains constant from generation to generation? The union of sperm and egg cells should *double* the number of chromosomes each generation. Yet this does not occur. Weismann inferred that the parent must have a process which reduces the chromosome number by half. Furthermore, this process has to occur in the germ plasm. Weismann's thinking on this was correct. The process of meiosis, worked out at the turn of the century, proves the alternation of diploid (somatic) and haploid (sexually reproductive) states of the cell.

The closing in on heredity through cytological studies was in process when a surprising discovery was made in 1900. Three independent investigations brought to attention an earlier finding about the inheritance of well-defined traits. Gregor Mendel, in 1859–1865, had used the garden pea, *Pisum sativum,* for an analysis of the inheritance of several different traits. His major discovery, the law of segregation, revealed an unexpected phenomenon. Sharply contrasted traits (different colors, different sizes) did not usually mix or blend but could be extracted in their original form in later generations. Mendel did not know what the hereditary factors were which carried these traits, but he did imply their existence.

Within a few years of the rediscovery of Mendel's work, cytologists had worked out a correspondence of the chromosome mechanics in meiosis and the law of segregation. This "chromosome theory of heredity" rapidly generated experiments, particularly among the students of Thomas Hunt Morgan at Columbia University. The fruit fly, *Drosophila melanogaster,* was used in Morgan's laboratory. Its rapid life cycle (two weeks), small size (one-eighth of an inch), and prolific reproduction (100 progeny per pair of parents) made it a useful organism for genetic studies.

The hereditary factors of Mendel now took on a more defined form. Morgan's group proved these factors to be parts of the chromosomes of the fly. The factors could be mapped and the chromosome was popularly compared to a string of beads. In 1910, these factors were called "genes," and by 1917 Morgan's group had popularized the chromosome theory as the theory of the gene.

The structure and function of the gene could not be determined until the 1950's. In the thirty-year period between the theory of the gene and its molecular analysis, geneticists, particularly H. J. Muller, demonstrated several of its important

features. The gene was shown to be stable; it can make tens of thousands of copies of itself with few errors while those rare errors or *mutations* which do occur do not affect the capacity of the gene to divide. The gene usually affects several characters, and it has been demonstrated that any one character is affected by tens or hundreds of genes. The theory of the gene provided an understanding of the missing clue to Darwin's theory—the origin of variation. The proof, by Muller, that genes could be mutated artificially led to numerous advances in evolution theory. Muller believed that the origin of life itself could be attributed to the first *gene,* not the first cell. The recognition of the importance of the gene led to the biochemical study of gene function. The gene was shown by G. W. Beadle and his colleagues to affect the production of individual enzymes. Very rapidly after his discovery genes were shown to consist of nucleic acid.

With the chemical basis of the gene established, a great interest in nucleic acids was initiated. In 1953 one of the most important biological discoveries of this century was made by J. D. Watson and F. H. C. Crick. They worked out a model for the structure of the nucleic acid associated with heredity, DNA (deoxyribonucleic acid). This model predicted the manner in which genes replicate; it predicted the chemical basis of mutation; and it predicted the chemical basis for a genetic code. The years since 1953 have witnessed a remarkable progression of chemical and physical studies of the gene and its function. No period of time except for the years 1910 to 1915 compares to it in the number of significant discoveries.

Gregor Mendel

Letter to Carl Nägeli

One of the remarkable events in the history of science is Gregor Mendel's long wait for fame. Mendel did not live to see that day in 1900 when his work was zealously championed and attacked by the world's greatest biologists. In 1865, when Mendel published his results in his local scientific society at Brünn, there was no immediate reaction. Many did not understand what his laws meant. For most biologists heredity dealt with minor fluctuating differences, virtually imperceptible in degree. Mendel, instead, had chosen grossly different characteristics and asserted a mathematical rule governing their distribution from the parents to their progeny. Perhaps the major reason for Mendel's failure to convince his generation is that *he didn't try to convince it!* If he had wished to do so, he could have sent copies to Darwin and Huxley, to Hooker, or Asa Gray. There were many prominent evolutionists well aware of the need for a theory of heredity. Mendel could also have attempted to publish his paper in a more prominent journal, or, like most of his contemporaries, he could have expanded his work to *book length* and then publish a provocative essay calling attention to the significance of his findings. It is likely that Mendel, an Augustinian monk, was by temperament and training too humble to stir up a debate of this magnitude.

In this letter to Nägeli, Mendel summarizes his work on peas and alludes to other work which was not as conclusive. Mendel's major contributions are today associated with his name. *The law of segregation* asserts that the hybrid formed by the union of reproductive cells carrying contrasting characteristics will return these characteristics unchanged to the next generation through the reproductive cells of the hybrid. *The law of independent assortment* asserts that several pairs of contrasting characters segregate independently of one another and their numbers may be

predicted mathematically. Mendel also noticed that the hybrid frequently expressed only *one* of the two contrasting traits that he selected for the parents. The expressed character in the hybrid Mendel called *dominant* and the unexpressed character he called *recessive*. The term "gene" would not be coined for forty-four more years, but Mendel's work implied the existence of what he called "formative elements" which brought about the contrasting characters he used in his experiments.

Highly Esteemed Sir:

My most cordial thanks for the printed matter you have so kindly sent me! The papers "die Bastardbildung im Pflanzenreiche," "über die abgeleiteten Pflanzenbastarde," "die Theorie der Bastardbildung," "die Zwischenformen zwischen den Pflanzenarten," "die systematische Behandlung der Hieracien rücksichlich der Mittelformen und des Umfangs der Species," especially capture my attention. This thorough revision of the theory of hybrids according to contemporary science was most welcome. Thank you again!

With respect to the essay which your honor had the kindness to accept, I think I should add the following information: the experiments which are discussed were conducted from 1856 to 1863. I knew that the results I obtained were not easily compatible with our contemporary scientific knowledge, and that under the circumstances publication of one such isolated experiment was doubly dangerous; dangerous for the experimenter and for the cause he represented. Thus I made every effort to verify, with other plants, the results obtained with Pisum. A number of hybridizations undertaken in 1863 and 1864 convinced me of the difficulty of finding plants suitable for an extended series of experiments, and that under unfavorable circumstances years might elapse without my obtaining the desired information. I attempted to inspire some control experiments, and for that reason discussed the Pisum experiments at the meeting of the local society of naturalists. I encountered, as was to be expected, divided opinion; however, as far as I know, no one undertook to repeat the experiments. When, last year, I was asked to publish my lecture in the proceedings of the society, I agreed to do so, after having re-examined my records for the various years of experimentation, and not having been able to find a source of error. The paper which was submitted to you is the unchanged

reprint of the draft of the lecture mentioned; thus the brevity of the exposition, as is essential for a public lecture.

I am not surprised to hear your honor speak of my experiments with mistrustful caution; I would not do otherwise in a similar case. Two points in your esteemed letter appear to be too important to be left unanswered. The first deals with the question whether one may conclude that constancy of type has been obtained if the hybrid Aa produces a plant A, and this plant in turn produces only A.

Permit me to state that, as an empirical worker, I must define constancy of type as the retention of a character during the period of observation. My statements that some of the progeny of hybrids breed true to type thus includes only those generations during which observations were made; it does not extend beyond them. For two generations all experiments were conducted with a fairly large number of plants. Starting with the third generation it became necessary to limit the numbers because of lack of space, so that, in each of the seven experiments, only a sample of those plants of the second generation (which either bred true or varied) could be observed further. The observations were extended over four to six generations (p. 13). Of the varieties which bred true (pp. 15–18) some plants were observed for four generations. I must further mention the case of a variety which bred true for six generations, although the parental types differed in four characters. In 1859 I obtained a very fertile descendent with large, tasty, seeds from a first generation hybrid. Since, in the following year, its progeny retained the desirable characteristics and were uniform, the variety was cultivated in our vegetable garden, and many plants were raised every year up to 1865. The parental plants were bcDg and BCdG:

B. albumen yellow	b. albumen green
C. seed-coat grayish-brown	c. seed-coat white
D. pod inflated	d. pod constricted
G. axis long	g. axis short

The hybrid just mentioned was BcDG.

The color of the albumen could be determined only in the plants saved for seed production, for the other pods were harvested in an immature condition. Never was green albumen observed in these plants, reddish-purple flower color (an indication of brown seed-coat), constriction of the pod, nor short axis.

This is the extent of my experience. I cannot judge whether these findings would permit a decision as to constancy of type; however, I am inclined to regard the separation of parental characteristics in the progeny of hybrids in Pisum as complete, and thus permanent. The progeny of hybrids carries one or the other of the parental characteristics, or the hybrid form of the two; I have never observed gradual transitions between the parental characters or a progressive approach toward one of them. The course of development consists simply in this; that in each generation the two parental characteristics appear, separated and unchanged, and there is nothing to indicate that one of them has either inherited or taken over anything from the other. For an example, permit me to point to the packets, numbers 1035–1088, which I sent you. All the seeds originated in the first generation of a hybrid in which brown and white seed-coats were combined. Out of the brown seed of this hybrid, some plants were obtained with seed-coats of a pure white color, without any admixture of brown. I expect those to retain the same constancy of character as found in the parental plant.

The second point, on which I wish to elaborate briefly, contains the following statement: "You should regard the numerical expressions as being only empirical, because they can not be proved rational."

My experiments with single characters all lead to the same result: that from the seeds of the hybrids, plants are obtained half of which in turn carry the hybrid character (Aa), the other half, however, receive the parental characters A and a in equal amounts. Thus, on the average, among four plants two have the hybrid character Aa, one the parental character A, and the other the parental character a. Therefore $2Aa + A + a$ or $A + 2Aa + a$ is the empirical simple, developmental series for two differentiating characters. Likewise it was shown in an empirical manner that, if two or three differentiating characters are combined in the hybrid, the developmental series is a combination of two or three simple series. Up to this point I don't believe I can be accused of having left the realm of experimentation. If then I extend this combination of simple series to any number of differences between the two parental plants, I have indeed entered the rational domain. This seems permissible, however, because I have proved by previous experiments that the development of any two differentiating characteristics proceeds independently of

any other differences. Finally, regarding my statements on the differences among the ovules and pollen cells of the hybrids; they also are based on experiments. These and similar experiments on the germ cells appear to be important, for I believe that their results furnish the explanation for the development of hybrids as observed in Pisum. These experiments should be repeated and verified.

I regret very much not being able to send your honor the varieties you desire. As I mentioned above, the experiments were conducted up to and including 1863; at that time they were terminated in order to obtain space and time for the growing of other experimental plants. Therefore seeds from those experiments are no longer available. Only one experiment on differences in the time of flowering was continued; and seeds are available from the 1864 harvest of this experiment. These are the last I collected, since I had to abandon the experiment in the following year because of devastation by the pea beetle, *Bruchus pisi*. In the early years of experimentation this insect was only rarely found on the plants, in 1864 it caused considerable damage, and appeared in such numbers in the following summer that hardly a 4th or 5th of the seeds was spared. In the last few years it has been necessary to discontinue cultivation of peas in the vicinity of Brünn. The seeds remaining can still be useful, among them are some varieties which I expect to remain constant; they are derived from hybrids in which two, three, and four differentiating characters are combined. All the seeds were obtained from members of the first generation, i.e., of such plants as were grown directly from the seeds of the original hybrids.

I should have scruples against complying with your honor's request to send these seeds for experimentation, were it not in such complete agreement with my own wishes. I fear that there has been partial loss of viability. Furthermore the seeds were obtained at a time when *Bruchus pisi* was already rampant, and I cannot acquit this beetle of possibly transferring pollen; also, I must mention again that the plants were destined for a study of differences in flowering time. The other differences were also taken into account at the harvest, but with less care than in the major experiment. The legend which I have added to the packet numbers on a separate sheet is a copy of the notes I made for each individual plant, with pencil, on its envelope at the time of harvest. The dominant characters are designated as A, B, C,

D, E, F, G and as concerns their dual meaning please refer to p. 11. The recessive characters are designated a, b, c, d, e, f, g; these should remain constant in the next generation. Therefore, from those seeds which stem from plants with recessive characters only, identical plants are expected (as regards the characters studied).

Please compare the numbers of the seed packets with those in my record, to detect any possible error in the designations—each packet contains the seeds of a single plant only.

Some of the varieties represented are suitable for experiments on the germ cells; their results can be obtained during the current summer. The round yellow seeds of packets 715, 730, 736, 741, 742, 745, 756, 757, and on the other hand, the green angular seeds of packets 712, 719, 734, 737, 749, and 750 can be recommended for this purpose. By repeated experiments it was proved that, if plants with green seeds are fertilized by those with yellow seeds, the albumen of the resulting seeds has lost the green color and has taken up the yellow color. The same is true for the shape of the seed. Plants with angular seeds, if fertilized by those with round or rounded seeds, produce round or rounded seeds. Thus, due to the changes induced in the color and shape of the seeds by fertilization with foreign pollen, it is possible to recognize the constitution of the fertilizing pollen.

Let B designate yellow color; b, green color of the albumen.

Let A designate round shape; a, angular shape of the seeds.

If flowers of such plants as produce green and angular seeds by self-fertilization are fertilized with foreign pollen, and if the seeds remain green and angular, then the pollen of the donor plant was, as regards the two characters .... ab

If the shape of the seeds is changed, the pollen was taken from .... Ab

If the color of the seeds is changed, the pollen was taken from .... aB

If both shape and color are changed, the pollen was taken from .... AB

The packets enumerated above contain round and yellow, round and green, angular and yellow, and angular and green seeds from the hybrids ab + AB. The round and yellow seeds would be best suited for the experiment. Among them (see experiment p. 15) the varieties AB, ABb, Aab, and AaBb may occur; thus four cases are possible when plants grown from green

and angular seeds are fertilized by the pollen of those grown from the above-mentioned round and yellow seeds, i.e.

I. ab + AB
II. ab + ABb
III. ab + AaB
IV. ab + AaBb

If the hypothesis that hybrids form as many types of pollen cells as there are possible constant combination types is correct, plants of the makeup

AB produce pollen of the type AB
ABb " " " " " AB and Ab
AaB " " " " " AB and aB
AaBb " " " " " AB, Ab, aB, and ab

Fertilization of ovules occurs:

I. Ovules ab with pollen AB
II. " ab " " AB and Ab
III. " ab " " AB and aB
IV. " ab " " AB, Ab, aB, and ab

The following varieties may be obtained from this fertilization:

I. AaBb
II. AaBb and Aab
III. AaBb and aBb
IV. AaBb, Aab, aBb, and ab

If the different types of pollen are produced in equal numbers, there should be in

I. All seeds round and yellow
II. one half round and yellow
one half round and green
III. one half round and yellow
one half angular and yellow
IV. one quarter round and yellow
one quarter round and green
one quarter angular and yellow
one quarter angular and green

Furthermore, since the numerical relations between AB, ABb, AaB, AaBb are 1:2:2:4, among any nine plants grown from round yellow seed there should be found on the average AaBb four times, ABb and AaB twice each, and AB once; thus the IVth case should occur four times as frequently as the Ist and twice as frequently as the IInd or IIIrd.

If, on the other hand, plants grown from the round yellow

seeds mentioned are fertilized by pollen from green angular plants, the results should be exactly the same, provided that the ovules are of the same types, and formed in the same proportions, as was reported for the pollen.

I have not performed this experiment myself, but I believe, on the basis of similar experiments, that one can depend on the result indicated.

In the same fashion individual experiments may be performed for each of the two seed characters separately, all those round seeds which occurred together with angular ones, and all the yellow ones which occurred with green seeds on the same plant are suitable. If, for instance, a plant with green seeds was fertilized by one with yellow seeds, the seeds obtained should be either 1) all yellow, or 2) half yellow and half green, since the plants originating from yellow seeds are of the varieties B and Bb. Since, furthermore, B and Bb occur in the ratio of 1:2, the 2nd fertilization will occur twice as frequently as the 1st.

Regarding the other characters, the experiments may be conducted in the same way; results, however, will not be obtained until next year.

I have all the piloselloid Hieracia which your honor recommends for the experiments; also *H. murorum* and *H. vulgatum* of the Archieracia; *H. glaucum, H. alpinum, H. amplexicaule, H. prenanthoides,* and *H. tridentatum* do not occur in this vicinity. Last summer I found a withered Hieracium, which has the seed color of Prenanthoidea (Fries: Achaenia typice testaces [pallida]), but did not resemble any of the herbarium specimens of this type very closely; finally our botanist declared it to be a hybrid. The rootstock has been transplanted to the garden for further observations, and the seeds have been planted. On the whole, this area is poor in Hieracia, and probably has not been sufficiently searched. Next summer I hope to have the time to roam the sandy lignite country which extends eastward from Brünn for several miles to the Hungarian frontier. Several other rare plants are known from this region. The Moravian plateau also is probably terra incognita as far as the Hieracia are concerned. If I should find anything noteworthy during the summer, I shall hurry to send it to your honor. At the moment permit me to include with the seed packets the plant just mentioned, albeit in a rather defective condition, together with another Hieracium. Last year I found at least 50 specimens of it on an old garden

wall. This plant is not found in the local herbaria; its appearance suggests both *H. praealtum* and *H. echioides,* without being one or the other. *H. praealtum* does occur in the environs of the city, *H. echioides* does not.

Several specimens of the hybrid *Geum urbanum* + *G. rivale* (from last year's hybridization) wintered in the greenhouse. Three are now flowering, the others will follow. Their pollen is fairly well developed, and the plants should be fertile, just as Gärtner states. It seems strange that all the plants now flowering are of the exceptional type mentioned by Gärtner. He says: "*Geum urbano-rivale,* mostly with large flowers, like *rivale,* and only a few specimens with small yellow flowers like *urbanum.*" In my plants the flowers are yellow or yellow-orange, and about half the size of those of *G. rivale;* the other characters correspond, as far as can be judged at present, to those of *G. intermedium* Ehrh. Could it be that the exceptional type has an earlier flowering season? To judge from the buds, however, the other plants do not have large flowers either. Or could it be that the exception has become the rule? I believe I have good reasons for considering my parental species pure. I obtained *G. urbanum* in the environs of the city, where neither *G. rivale* nor any other species of the genus occurs; and I got *G. rivale* in a damp mountain meadow, where *G. urbanum* certainly does not occur. This plant has all the characteristics of *G. rivale;* it is being maintained in the garden, and seedlings have been produced from self-fertilization.

The *Cirsium arvense* + *C. oleraceum* hybrids, sown in the fall, have died during the winter; one plant of the *C. arvense* + *C. canum* hybrid survived. I hope the spring seedlings will do better. Two other Cirsium hybrids have wintered well in the greenhouse. Last summer I observed, on a flowering plant of *C. praemorsum* M. (*olerac.* + *rivulare*), that in those heads which develop first and last on the stems, no pollen is formed, and thus they are completely sterile; on the others (about one half of the total heads) some pollen and fertile seed is formed. Fertilization experiments were conducted with two of the late-developing heads; pollen of *C. palustre* was transferred to one, pollen of *C. canum* to the other. Viable seeds were obtained from both, the resulting plants survived the winter in the greenhouse, and are now developed to a stage at which the success of the hybridization is evident. Some seedlings of *C. praemorsum,* others of a hybrid

(probably in the group *C. canum* + *palustre*), and those of a third one, probably *C. rivulare* + *palustre,* have survived the winter in the open quite well. The same may be said of the autumn seedlings of the hybrids *Aquilegia canadensis* + *vulgaris, A. canadensis* + *A. atropurpurea,* and *A. canadensis* + *A. Wittmaniana.* However, fall seedlings of some Hieracia which were grown to test constancy of type have suffered considerable damage. In this genus it is preferable to sow in the early spring, but then it is doubtful that the plants will flower in the same year. Nevertheless, Fries has made this statement concerning the division Accipitrina: "Accipitrina, praecocius sata, vulgo primo anno florent."

I have obtained luxuriant plants of *Linaria vulgaris* + *L. purpurea;* I hope they will flower in the first year. The same may be said of *Calceolaria salicifolia* and *C. rugosa.* Hybrids of *Zea Mays major* (with dark red seeds) + *Z. Mays minor* (with yellow seeds) and of *Zea Mays major* (with dark red seeds) + *Zea Cuzko* (with white seeds) will develop during the summer. Whether *Zea Cuzko* is a true species or not I do not dare to state. I obtained it with this designation from a seed dealer. At any rate it is a very aberrant form. To study color development in flowers of hybrids, cross-fertilizations were made last year between varieties of *Ipomoea purpurea, Cheiranthus annuus,* and *Antirrhinum majus.* An experiment with hybrids of *Tropaeoleum majus* + *T. minus* (1st generation) must also be mentioned.

For the current year exploratory experiments with Veronica, Viola, Potentilla, and Carex are planned. Unfortunately, I have only a small number of species.

Because of lack of space the experiments can be started with a small number of plants only; after the fertility of the hybrids has been tested, and when it is possible to protect them sufficiently during the flowering period, each in turn will receive an extensive study. Thus far the three Aquilegia hybrids mentioned above and *Tropaeoleum majus* + *T. minus* are suitable, although the latter has only partial fertility. It is hoped that *Geum urbanum* + *G. rivale* can be included in the group of suitable plants.

As must be expected, the experiments proceed slowly. At first beginning, some patience is required, but later, when several experiments are progressing concurrently, matters are improved. Every day, from spring to fall, one's interest is refreshed daily, and the care which must be given to one's wards is thus amply

repaid. In addition, if I should, by my experiments, succeed in hastening the solution of these problems, I should be doubly happy.

Accept, highly esteemed Sir, the expression of most sincere respect from

Your devoted,
G. MENDEL
(Altbrünn, Monastery of St. Thomas)

Brünn, 18 April, 1867

William Bateson

Problems of Heredity as a Subject for Horticultural Investigation

William Bateson began his career as an embryologist, studying at Cambridge in the 1870's. He was introduced to heredity as a field of study by the American biologist W. K. Brooks. Toward the close of the nineteenth century, Bateson argued strongly for evolution by discontinuous variation. Bateson hoped to convince naturalists, collectors, and breeders that they should record *all* progeny and *all* collected specimens rather than the "best" representatives. In 1899 Bateson initiated a series of experiments which revealed the discontinuous variation of inherited traits. To stimulate further research he presented these results to the Royal Horticultural Society. Bateson prepared the second of these papers early in 1900. While traveling on the train to the Society meetings, he encountered among his mail Hugo De Vries's account of Mendel's laws. Bateson was so struck by the significance of Mendel's work that he revised the paper and presented this first English account of one of the greatest experiments in science.

Bateson's enthusiasm was not shared by the majority of British evolutionists. They interpreted Darwinism as incompatible with Mendelism. It was Bateson's greatest contribution to genetics that he took on all his critics and eventually forced them to acknowledge that Mendelism was here to stay.

An exact determination of the laws of heredity will probably work more change in man's outlook on the world, and in his power over nature, than any other advance in natural knowledge that can be foreseen.

There is no doubt whatever that these laws can be determined. In comparison with the labour that has been needed for other great discoveries it is even likely that the necessary effort will be

Reprinted from *The Journal of the Royal Horticultural Society XXV*:54–61, 1900, by permission of the Society.

small. It is rather remarkable that while in other branches of physiology such great progress has of late been made, our knowledge of the phenomena of heredity has increased but little; though that these phenomena constitute the basis of all evolutionary science and the very central problem of natural history is admitted by all. Nor is this due to the special difficulty of such inquiries so much as to general neglect of the subject.

It is in the hope of inducing others to pursue these lines of investigation that I take the problems of heredity as the subject of this lecture to the Royal Horticultural Society.

No one has better opportunities of pursuing such work than horticulturists. They are daily witnesses of the phenomena of heredity. Their success depends also largely on a knowledge of its laws, and obviously every increase in that knowledge is of direct and special importance to them.

The want of systematic study of heredity is due chiefly to misapprehension. It is supposed that such work requires a lifetime. But though for adequate study of the complex phenomena of inheritance long periods of time must be necessary, yet in our present state of deep ignorance almost of the outline of the facts, observations carefully planned and faithfully carried out for even a few years may produce results of great value. In fact, by far the most appreciable and definite additions to our knowledge of these matters have been thus obtained.

There is besides some misapprehension as to the kind of knowledge which is especially wanted at this time, and as to the modes by which we may expect to obtain it. The present paper is written in the hope that it may in some degree help to clear the ground of these difficulties by a preliminary consideration of the question, How far have we got towards an exact knowledge of heredity, and how can we get further?

Now this is pre-eminently a subject in which we must distinguish what we *can* do from what we want to do. We *want* to know the whole truth of the matter; we want to know the physical basis, the inward and essential nature, "the causes", as they are sometimes called, of heredity. We want also to know the laws which the outward and visible phenomena obey.

Let us recognise from the outset that as to the essential nature of these phenomena we still know absolutely nothing. We have no glimmering of an idea as to what constitutes the essential process by which the likeness of the parent is transmitted to the

offspring. We can study the processes of fertilisation and development in the finest detail which the microscope manifests to us, and we may fairly say that we have now a thorough grasp of the visible phenomena; but of the nature of the physical basis of heredity we have no conception at all. No one has yet any suggestion, working hypothesis, or mental picture that has thus far helped in the slightest degree to penetrate beyond what we see. The process is as utterly mysterious to us as a flash of lightning is to a savage. We do not know what is the essential agent in the transmission of parental characters, not even whether it is a material agent or not. Not only is our ignorance complete, but no one has the remotest idea how to set to work on that part of the problem. We are in the state in which the students of physical science were in the period when it was open to anyone to believe that heat was a material substance or not, as he chose.

But apart from any conception of the essential modes of transmission of characters, we can study the outward facts of the transmission. Here, if our knowledge is still very vague, we are at least beginning to see how we ought to go to work. Formerly naturalists were content with the collection of numbers of isolated instances of transmission—more especially, striking and peculiar cases—the sudden appearance of highly prepotent forms, and the like. We are now passing out of that stage. It is not that the interest of particular cases has in any way diminished—for such records will always have their value—but it has become likely that general expressions will be found capable of sufficiently wide application to be justly called "laws" of heredity. That this is so is due almost entirely to the work of Mr. F. Galton, to whom we are indebted for the first systematic attempt to enunciate such a law.

All the laws of heredity so far propounded are of a statistical character and have been obtained by statistical methods. If we consider for a moment what is actually meant by a "law of heredity", we shall see at once why these investigations must follow statistical methods. For a "law" of heredity is simply an attempt to declare the course of heredity under given conditions. But if we attempt to predicate the course of heredity we have to deal with conditions and groups of causes wholly unknown to us, whose presence we cannot recognise, and whose magnitude we

cannot estimate in any particular case. The course of heredity in particular cases therefore cannot be foreseen.

Of the many factors which determine the degree to which a given character shall be present in a given individual only one is known to us, namely, the degree to which the character is present in the parents. It is common knowledge that there is not that close correspondence between parent and offspring which would result were this factor the only one operating; but that, on the contrary, the resemblance between the two is only a general one.

In dealing with phenomena of this class the study of single instances reveals no regularity. It is only by collection of facts in great numbers, and by statistical treatment of the mass, that any order or law can be perceived. In the case of a chemical reaction, for instance, by suitable means the conditions can be accurately reproduced, so that in every individual case we can predict with certainty that the same result will occur. But with heredity it is somewhat as it is in the case of the rainfall. No one can say how much rain will fall to-morrow in a given place, but we can predict with moderate accuracy how much will fall next year, and for a period of years a prediction can be made which accords very closely with the truth.

Similar predictions can from statistical data be made as to the duration of life and a great variety of events the conditioning causes of which are very imperfectly understood. It is predictions of this kind that the study of heredity is beginning to make possible, and in that sense laws of heredity can be perceived.

We are as far as ever from knowing why some characters are transmitted, while others are not; nor can anyone yet foretell which individual parent will transmit characters to the offspring, and which will not; nevertheless the progress made is distinct.

As yet investigations of this kind have been made in only a few instances, the most notable being those of Galton on human stature, and on the transmission of colours in Basset hounds. In each of these cases he has shewn that the expectation of inheritance is such that a simple arithmetical rule is approximately followed. The rule thus arrived at is that of the whole heritage of the offspring the two parents together on an average contribute one-half, the four grandparents one-quarter, the eight great-grandparents one-eighth, and so on, the remainder being contributed by the remoter ancestors.

Such a law is obviously of practical importance. In any case to which it applies we ought thus to be able to predict the degree with which the purity of a strain may be increased by selection in each successive generation.

To take a perhaps impossibly crude example, if a seedling shew any particular character which it is desired to fix, on the assumption that successive self-fertilisations are possible, according to Galton's law the expectation of purity should be in the first generation of self-fertilisation 1 in 2, in the second generation 3 in 4, in the third 7 in 8, and so on.

But already many cases are known to which the rule in the simple form will not apply. Galton points out that it takes no account of individual prepotencies. There are, besides, numerous cases in which on crossing two varieties the character of one variety is almost always transmitted to the first generation. Examples of these will be familiar to those who have experience in such matters. The offspring of the Polled Angus cow and the Shorthorn bull is almost invariably polled. Seedlings raised by crossing *Atropa belladonna* with the yellow-fruited variety have without exception the blackish purple fruits of the type. In several hairy species when a cross with a glabrous variety is made, the first cross-bred generation is altogether hairy.

Still more numerous are examples in which the characters of one variety very largely, though not exclusively, predominate in the offspring.

These large classes of exceptions—to go no further—indicate that, as we might in any case expect, the principle is not of universal application, and will need various modifications if it is to be extended to more complex cases of inheritance of varietal characters. No more useful work can be imagined than a systematic determination of the precise "law of heredity" in numbers of particular cases.

Until lately the work which Galton accomplished stood almost alone in this field, but quite recently remarkable additions to our knowledge of these questions have been made. In the present year Professor De Vries published a brief account[1] of experiments which he has for several years been carrying on, giving results of the highest value.

The description is very short, and there are several points as to which more precise information is necessary both as to details of procedure and as to statement of results.[2] Nevertheless it is

impossible to doubt that the work as a whole constitutes a marked step forward, and the full publication which is promised will be awaited with great interest.

The work relates to the course of heredity in cases where definite varieties differing from each other in some *one* definite character are crossed together. The cases are all examples of discontinuous variation: that is to say, cases in which actual intermediates between the parent forms are not usually produced on crossing. It is shewn that the subsequent posterity obtained by self-fertilising these cross-breds or hybrids break up into the original parent forms according to fixed numerical rule.

Professor De Vries begins by reference to a remarkable memoir by Gregor Mendel,[3] giving the results of his experiments in crossing varieties of *Pisum sativum*. These experiments of Mendel's were carried out in a large scale, his account of them is excellent and complete, and the principles which he was able to deduce from them will certainly play a conspicuous part in all future discussions of evolutionary problems. It is not a little remarkable that Mendel's work should have escaped notice, and been so long forgotten.

For the purposes of his experiments Mendel selected seven pairs of characters as follows:

1. Shape of ripe seed, whether round, or angular and wrinkled.
2. Colour of "endosperm"[4] (cotyledons), whether some shade of yellow, or a more or less intense green.
3. Colour of the seed-skin, whether various shades of grey and grey-brown, or white.
4. Shape of seed-pod, whether simply inflated, or deeply constricted between the seeds.
5. Colour of unripe pod, whether a shade of green, or bright yellow.
6. Shape of inflorescence, whether the flowers are arranged along one axis, or are terminal and more or less umbellate.
7. Length of peduncle, whether about 6 or 7 inches long, or about ¾ to 1½ inches.

Large numbers of crosses were made between Peas differing in respect of each of these pairs of characters. It was found that in each case the offspring of the cross exhibited the character of one of the parents in almost undiminished intensity, and intermediates which could not be at once referred to one or other of the parental forms were not found.

In the case of each pair of characters there is thus one which in the first cross prevails to the exclusion of the other. This prevailing character Mendel calls the *dominant* character, the other being the *recessive* character.[4]

That the existence of such "dominant" and "recessive" characters is a frequent phenomenon in cross-breeding, is well known to all who have attended to these subjects.

By self-fertilising the cross-breds Mendel next raised another generation. In this generation were individuals which shewed the dominant character, but also individuals which preserved the recessive character. This fact also is known in a good many instances. But Mendel discovered that in this generation the numerical proportion of dominants to recessives is approximately constant, being in fact *as three to one*. With very considerable regularity these numbers were approached in the case of each of his pairs of characters.

There are thus in the first generation raised from the cross-breds 75 per cent. dominants and 25 per cent. recessives.

These plants were again self-fertilised, and the offspring of each plant separately sown. It next appeared that the offspring of the recessives *remained pure recessive,* and in subsequent generations never reverted to the dominant again.

But when the seeds obtained by self-fertilising the dominants were sown it was found that some of the dominants gave rise to pure dominants, while others had a mixed offspring, composed partly of recessives, partly of dominants. Here also it was found that the average numerical proportions were constant, those with pure dominant offspring being to those with mixed offspring as one to two. Hence it is seen that the 75 per cent. dominants really are not all alike, but consist of twenty-five which are pure dominants and fifty which are really cross-breds, though, like the cross-breds raised by crossing the two varieties, they only exhibit the dominant character.

To resume, then, it was found that by self-fertilising the original cross-breds the same proportion was always approached, namely:

25 dominants, 50 cross-breds, 25 recessives, or 1 D: 2 DR: 1 R.

Like the pure recessives, the pure dominants are thenceforth pure, and only give rise to dominants in all succeeding generations.

On the contrary the fifty cross-breds, as stated above, have mixed offspring. But these, again, in their numerical proportions,

follow the same law, namely, that there are three dominants to one recessive. The recessives are pure like those of the last generation, but the dominants can, by further self-fertilisation and cultivation of the seeds produced, be shewn to be made up of pure dominants and cross-breds in the same proportion of one dominant to two cross-breds.

The process of breaking up into the parent forms is thus continued in each successive generation, the same numerical law being followed so far as has yet been observed.

Mendel made further experiments with *Pisum sativum*, crossing pairs of varieties which differed from each other in *two* characters, and the results, though necessarily much more complex, shewed that the law exhibited in the simpler case of pairs differing in respect of one character operated here also.

Professor De Vries has worked at the same problem in some dozen species belonging to several genera, using pairs of varieties characterised by a great number of characters: for instance, colour of flowers, stems, or fruits, hairiness, length of style, and so forth. He states that in all these cases Mendel's law is followed.

The numbers with which Mendel worked, though large, were not large enough to give really smooth results; but with a few rather marked exceptions the observations are remarkably consistent, and the approximation to the numbers demanded by the law is greatest in those cases where the largest numbers were used. When we consider, besides, that Tschermak and Correns announce definite confirmation in the case of *Pisum*, and De Vries adds the evidence of his long series of observations on other species and orders, there can be no doubt that Mendel's law is a substantial reality; though whether some of the cases that depart mostly widely from it can be brought within the terms of the same principle or not, can only be decided by further experiments.

One may naturally ask, How can these results be brought into harmony with the facts of hybridisation as hitherto known; and, if all this is true, how is it that others who have so long studied the phenomena of hybridisation have not long ago perceived this law? The answer to this question is given by Mendel at some length, and it is, I think, satisfactory. He admits from the first that there are undoubtedly cases of hybrids and cross-breds which maintain themselves pure and do not break up. Such examples are plainly outside the scope of his law. Next he points out, what to anyone who has rightly comprehended the nature of discon-

tinuity in variation is well known, that the variations in *each* character must be *separately* regarded. In most experiments in crossing, forms are taken which differ from each other in a multitude of characters—some continuous, others discontinuous, some capable of blending with their contraries, while others are not. The observer on attempting to perceive any regularity is confused by the complications thus introduced. Mendel's law, as he fairly says, could only appear in such cases by the use of overwhelming numbers, which are beyond the possibilities of practical experiment.

Both these answers should be acceptable to those who have studied the facts of variation and have appreciated the nature of species in the light of those facts. That different species should follow different laws, and that the same law should not apply to all characters alike, is exactly what we have every right to expect. It will also be remembered that the principle is only declared to apply to discontinuous characters. As stated also it can only be true where reciprocal crossings lead to the same result. Moreover, it can only be tested when there is no sensible diminution in fertility on crossing.

Upon the appearance of De Vries' papers announcing the "rediscovery" and confirmation of Mendel's law and its extension to a great number of cases two other observers came forward and independently described series of experiments fully confirming Mendel's work. Of these papers the first is that of Correns,[5] who repeated Mendel's original experiment with Peas having seeds of different colours. The second is a long and very valuable memoir of Tschermak,[6] which gives an account of elaborate researches into the results of crossing a number of varieties of *Pisum sativum*. These experiments were in many cases carried out on a large scale, and prove the main fact enunciated by Mendel beyond any possibility of contradiction. Both Correns (in regard to maize) and Tschermak in the case of *P. sativum* have obtained further proof that Mendel's law holds as well in the case of varieties differing from each other in *two* characters, one of each being dominant, though of course a more complicated expression is needed in such cases.[7]

That we are in the presence of a new principle of the highest importance is, I think, manifest. To what further conclusions it may lead us cannot yet be foretold. But both Mendel and the authors who have followed him lay stress on one conclusion,

which will at once suggest itself to anyone who reflects on the facts. For it will be seen that the results are such as we might expect if it is imagined that the cross-bred plant produced pollen grains and ovules, each of which bears only *one* of the alternative varietal characters and not both. If this were so, and if on the average the same number of pollen grains and ovules partook of each of the two characters, it is clear that on a random assortment of pollen grain and ovules Mendel's law would be obeyed. For 25 per cent. of "dominant" pollen grains would unite with 25 per cent. "dominant" ovules; 25 per cent. "recessive" pollen grains would similarly unite with 25 per cent. "recessive" ovules; while the remaining 50 per cent. of each kind would unite together. It is this consideration which leads both De Vries and Mendel to assert that these facts of crossing prove that each ovule and each pollen grain is pure in respect of each character to which the law applies. It is highly desirable that varieties differing in the form of their pollen should be made the subject of these experiments, for it is quite possible that in such a case strong confirmation of this deduction might be obtained.

As an objection to this deduction, however, it is to be noted that though true intermediates did not occur, yet the degrees in which the characters appeared did vary, and it is not easy to see how the hypothesis of perfect purity in the reproductive cells can be supported in such cases. Be this, however, as it may, there is no doubt we are beginning to get new lights of a most valuable kind on the nature of heredity and the laws which it obeys. It is to be hoped that these indications will be at once followed up by independent workers. Enough has been said to shew how necessary it is that the subjects of experiment should be chosen in such a way as to bring the laws of heredity to a real test. For this purpose the first essential is that the differentiating characters should be few, and that all avoidable complications should be got rid of. Each experiment should be reduced to its simplest possible limits. The results obtained by Galton, and also the new ones especially detailed in this paper, have each been reached by restricting the range of observation to one character or group of characters, and there is every hope that by similar treatment our knowledge of heredity may be rapidly extended.

(NOTE. Since the above was printed further papers on Mendel's law have appeared, namely, De Vries, *Rev. gener. Bot.* 1900,

p. 257; Correns, *Bot. Ztg.* 1900, p. 229; and *Bot. Cblt.* LXXXIV, p. 97, containing new matter of importance. Professor De Vries kindly writes to me that in asserting the general applicability of Mendel's law to "monohybrids" (crosses between parents differing in respect of *one* character only), he intends to include cases of discontinuous varieties only, and he does not mean to refer to continuous varieties at all. 31 October 1900.)

1900

NOTES

[1] *Comptes Rendus,* 26 March 1900, and *Ber. d. Deutsch. Bot. Ges.* XVIII, 1900, p. 83.

[2] For example, I do not understand in what sense De Vries considers that Mendel's law can be supposed to apply even to all "monohybrids", for numerous cases are already known in which no such rule is obeyed.

[3] "Versuche üb. Pflanzenhybriden" in the *Verh. d. Naturf. Ver. Brünn* IV, 1865.

[4] Note that by these useful terms the complications involved by use of the expression "prepotent" are avoided.

[5] *Ber. deut. Bot. Ges.* 1900, XVIII, 158.

[6] *Zeitschr. f. d. landw. Versuchswesen in Oesterr.* 1900, III, 465.

[7] Tschermak's investigations were besides directed to a re-examination of the question of the absence of beneficial results on cross-fertilising *P. sativum,* a subject already much investigated by Darwin, and upon this matter also important further evidence is given in great detail.

T. H. Morgan

Random Segregation versus Coupling in Mendelian Inheritance

When Morgan first took a critical interest in Mendelism, he rejected the view that the factors (or *genes* as they were called after 1910) were associated with chromosomes. He reasoned that there were more traits than chromosomes and that the chromosome theory in this simplified form was inadequate. This reluctance to associate genes with chromosomes was abandoned in 1911. Morgan realized that two genes, both associated with sex determination, were both carried by the sex (X) chromosome. Furthermore, they could recombine with one another in all possible combinations. This phenomenon he attributed to the physical exchange (or "crossing over") of chromosome parts during germ-cell formation. His theory simplified and eventually replaced more cumbersome models for this phenomenon which were proposed by Bateson. Morgan inferred that there was a relation of distance between the genes and the frequency of crossing over, but the actual mathematical analysis of this process was worked out by his students.

Mendel's Law of inheritance rests on the assumption of random segregation of the factors for unit characters. The typical proportions for two or more characters, such as 9:3:3:1, etc., that characterize Mendelian inheritance, depend on an assumption of this kind. In recent years a number of cases have come to light in which when two or more characters are involved the proportions do not accord with Mendel's assumption of random segregation. The most notable cases of this sort are found in sex-limited inheritance in *Abraxas* and *Drosophila,* and in several breeds of poultry, in which a coupling between the factors for femaleness

Reprinted from *Science 34*:384, 1911, by permission of the publisher.

and one other factor must be assumed to take place, and in the case of peas where color and shape of pollen are involved. In addition to these cases Bateson and his collaborators (Punnett, DeVilmorin and Gregory) have recently published a number of new ones.

In order to account for the results Bateson assumes not only coupling, but also repulsions in the germ cells. The facts appear to be exactly comparable to those that I have discovered in *Drosophila,* and since these results have led me to a very simple interpretation, I venture to contrast Bateson's hypothesis with the one that I have to offer.

The facts on which Bateson bases his interpretation may be briefly stated in his own words, namely: "that if A, a and B, b, are two allelomorphic pairs subject to coupling and repulsion, the factors A and B will repel each other in the gametogenesis of the double heterozygote resulting from the union Ab × aB, but will be coupled in the gametogenesis of the double heterozygote resulting from the union AB × ab," and further, "We have as yet no probable surmise to offer as to the essential nature of this distinction, and all that can yet be said is that in these special cases the distribution of the characters in the heterozygote is affected by the distribution in the original pure parents." Bateson further points out that since "sex in the fowls act as a repeller of at least three other factors, . . . some of them may be found able to take precedence of the others in such a way as to annul the present repulsion with subsequent coupling as a consequence."

In place of attractions, repulsions and orders of precedence, and the elaborate systems of coupling, I venture to suggest a comparatively simple explanation based on results of inheritance of eye color, body color, wing mutations and the sex factor for femaleness in *Drosophila.* If the materials that represent these factors are contained in the chromosomes, and if those factors that "couple" be near together in a linear series, then when the parental pairs (in the heterozygote) conjugate like regions will stand opposed. There is good evidence to support the view that during the strepsinema stage homologous chromosomes twist around each other, but when the chromosomes separate (split) the split is in a single plane, as maintained by Janssens. In consequence, the original materials will, for short distances, be more likely to fall on the same side of the split, while remoter regions will be as likely to fall on the same side as the last, as on

the opposite side. In consequence, we find coupling in certain characters, and little or no evidence at all of coupling in other characters; the difference depending on the linear distance apart of the chromosomal materials that represent the factors. Such an explanation will account for all of the many phenomena that I have observed and will explain equally, I think the other cases so far described. The results are a simple mechanical result of the location of the materials in the chromosomes and of the method of union of homologous chromosomes, and the proportions that result are not so much the expression of a numerical system as of the relative location of the factors in the chromosomes. Instead of random segregation in Mendel's sense we find "association of factors" that are located near together in the chromosomes. Cytology furnishes the mechanism that the experimental evidence demands.

1911

H. J. Muller

Mutation

Muller clarified the concept of mutation in 1923 by restricting the term to its present connotation—a change in the individual gene. The fourteen points listed by Muller include several major points frequently overlooked: the gene is extremely stable; its rare mutations occur almost randomly (they *cannot be directed* by applied agents or the environment); mutant genes retain the capacity to reproduce and the copied genes express the mutant condition; and the mutational event is highly localized not only within the cell but within only one chromosome of a cell.

The recognition that mutation and evolution are inseparable is emphasized by Muller. He also projects the implications of mutation for man and advocates consideration of future eugenic programs to maintain the human genetic heritage.

Beneath the imposing building called "Heredity" there has been a dingy basement called "Mutation." Lately the searchlight of genetic analysis has thrown a flood of illumination into many of the dark recesses there, revealing some of them as ordinary rooms in no wise different from those upstairs, that merely need to have their blinds flung back, while others are seen to be subterranean passageways of quite a different type. In other words, the term "mutation" originally included a number of distinct phenomena, which, from a genetic point of view, have nothing in common with one another. They were classed together merely because they all involved the sudden appearance of a new genetic type. Some have been found to be special cases of Mendelian recombination, some to be due to abnormalities in the distribution of entire chromosomes, and others to consist in changes in the individual genes or hereditary units. It seems incumbent upon us, however, in

Reprinted from *Eugenics, Genetics and the Family 1*:106–112, 1923, by permission of the publisher and the author.

the interests of scientific clarity, to agree to confine our use of the term "mutation" to one coherent class of events. The usage most serviceable for our modern purpose would be to limit the meaning of the term to the cases of the third type—that is, to real changes in the gene. This would also be most in conformity with the spirit of the original usage, for even in the earlier days, mutations were conceived of as fundamental changes in the hereditary constitution, and there were never intentionally included among them cases merely involving redistribution of hereditary units—when these cases were recognizable as such. In accordance with these considerations, our new definition would be: "mutation is alteration of the gene." And "alteration," as here used, is of course understood to mean a change of a transmissible, or at least of a propagable, sort.

In thus trimming down the scope of our category of mutation we do not deprive it of the material of most fundamental evolutionary significance. For all changes due to the redistribution of individual genes or of groups of genes, into new combinations, proportions, or quantities, are obviously made possible only by the prior changes that make these genes differ from each other in the first place. It should in addition be noted that changes due merely to differences in the gross proportions of entire groups of genes must be relatively incapable of that delicate adjustment which is required for evolutionary adaptation. And as to the question, frequently raised, whether all evolution is ultimately due to mutation, this is necessarily answered in the affirmative by our definitions of the gene and of mutation, which designate the gene as any unit of heredity, and mutation as any transmissible change occurring in the gene. The question of the basic mechanism of evolution thus becomes transferred to the problem of the character, frequency, and mode of occurrence of mutation, taken in this precise, yet comprehensive sense. And since eugenics is a special branch of evolutionary science it must be equally concerned with this problem.

In choosing the body of data wherewith to attack these questions of mutation, in their new form, it must unfortunately be recognized that the results with the evening primrose, Oenothera, although they formed the backbone of the earlier mutation theory, can no longer be regarded as having a direct bearing on the modern problem, since they cannot be shown to be due directly to changes in the genes. Certain of them, such as gigas, lata,

scintillans, etc., have been proved by Geerts, Lutz, Gates, and others, to be due to abnormalities in the apportionment of the chromosomes. Very valuable information on the genetics of cases of this sort is now being obtained, especially in the work of Blakeslee, Belling, and Farnham on much clearer cases of similar character in the Jimson weed, and, finally, in the work of Bridges on the fruit fly Drosophila. Most of the other so-called mutations in the evening primrose appear to be due to the normal hereditary processes of segregation and crossing over, working on a genetic constitution of a special type. Evidence for this was obtained in my analysis of the analogous case existing in the fly Drosophila, as follows. It had previously been shown by De Vries, and further elaborated by Renner, that germ cells or individuals of Oenothera bearing certain genes always died, in such a way that all the surviving individuals were heterozygous (hybrid) in regard to these genes. I later showed, through work on Drosophila, that when such a condition (there called "balanced lethal factors") exists, the situation tends to become still further complicated through the presence of other heterozygous genes, which are linked to those which cause death. When one or a group of these non-lethal genes crosses over (separates) from the lethals, as they occasionally do, they may become homozygous, producing a visible effect. Thus new types of individuals appear which may be ascribed to "mutation," whereas they are really due to crossing over. The work of Frost on stocks has shown that a precisely analogous situation exists in that form also, and G. H. Shull is obtaining direct evidence for the same conclusion in the evening primrose itself. In any event, it must be granted that so long as this interpretation cannot be definitely refuted, these variations cannot be used as examples on which to base our theory of gene change. In place, then, of the elaborate system of conclusions which has derived its support chiefly from the results in the evening primrose, it will be necessary for our present theory of gene change to erect an independent structure, built upon an entirely new basis.

The data upon which the new theory must be built consist of two main sorts, which may be called direct and indirect. (1) In the cases giving the direct evidence, the occurrence of the gene change can be proved, and it is possible to exclude definitely all alternative explanations, such as contamination of the material, emergence of previously "latent" factors, non-disjunction, etc.

So far, the only considerable body of such evidence is that gotten in the Drosophila work, where mutations have (in this sense) been actually observed in at least 100 loci. Considered collectively, however, there exist in other organisms enough scattered data to afford ample corroborative evidence for the generality of occurrence of mutations like those observed in the Drosophila work. In addition several specially mutable genes have been found in a number of plants (as well as in Drosophila) that are giving highly valuable information along their particular lines. And a number of selection experiments that have been performed on non-segregating lines of various organisms have also given us direct evidence, if not of the frequency, then at least of the infrequency, of mutations. (2) As for the indirect data, these may be gotten by examination of Mendelian factor-differences of all kinds, on the assumption that they must have arisen through mutation. Although this assumption can be shown to be fully justified, these cases cannot provide information concerning the manner of origin of the mutants, nor can they furnish a reliable index of the frequency of mutations, since the mutant genes may have been subjected to an unknown amount of selective elimination or selective propagation before the observations were taken. As for the still more indirect data, derived from studies of phylogenetic series and comparisons between different species, genera, etc., these occasionally give suggestive results, but where crosses cannot be made or where the differences cannot be traced down to the individual genes, such facts can seldom lead to trustworthy genetic conclusions.

On these various data, duly weighted, we may found our new mutation theory. We know nothing, as yet, about the mechanism of mutation, or about the nature of the gene—aside from the fact that nearly all genes hitherto studied behave like material particles existing in the chromosomes. Nevertheless there is already evidence for a number of empirical principles regarding the changes of the genes, some of which may conveniently be listed here in the form of 14 statements. I shall have opportunity merely to present these principles, without attempting any adequate explanations of how they have been derived from the data.

1. The first and probably most important principle is that most genes—both mutant and "normal"—are exceedingly stable. Some idea of the degree of this stability may be obtained from some quantitative studies of mutation which Altenburg and I have

made in the fruit fly Drosophila. It may be calculated from these experiments that a large proportion of the genes in Drosophila must have a stability which—at a minimum value—is comparable with that of radium atoms. Radium atoms, it may be recalled, have a so-called "mean life" of about two thousand years.

2. Certain genes are, however, vastly more mutable than others. For example, a gene causing variegation in corn, studied by Emerson, and another in the four-o'clock, studied by Maryatt, ordinarily have a mean life of only a few years; and that causing bar eye in Drosophila has a mean life of only about 65 years, as is shown by the results of Zeleny. (In expressing these results we are here using the physicists' index of stability, which seems most appropriate for the present purpose also.)

3. External agents do not ordinarily increase the mutability sufficiently (if at all) to cause an obvious "production" of mutation.

4. The changes are not exclusively of the character of losses; this is shown by the well-established occurrence of reverse mutations, in bar-eyed and white-eyed Drosophila, in Blakeslee's dwarf Portulaca, Emerson's variegated corn, and probably in a number of other recorded instances. It is known that mutations having an effect similar to that of losses do occur, however, and they may be relatively frequent.

5. The change in a given gene is not in all cases in the same direction, and it does not even, in all cases, involve the same characters. The latter point is illustrated by a series of mutations which I am investigating in Drosophila, which all involve one gene, but which produce, as the case may be, either a shortened wing, an eruption on the thorax, a lethal effect, or any combination of these three.

6. The direction of mutation in a given gene is, however, preferential, occurring oftener in some directions than in others. This is well illustrated in the studies on variegated corn and four-o'clocks, and on the bar eye and white eye and other series in Drosophila.

7. The mutability and preferential direction may themselves become changed through mutation, as illustrated by some of the same cases.

8. The mutations do not ordinarily occur in two or more different genes at once. In only two instances in Drosophila have mutations been found in two different, separated[1] genes in the

same line of cells of one individual. But a recurrent case, apparently of this kind, has recently been described in oats, by Nilsson-Ehle.

9. Not only does the mutation usually involve but one kind of gene—it usually involves but one gene of that kind in the cell. That is, the allelomorphs mutate independently of one another, just as totally different genes do. There is evidence for this derived from corn, Portulaca, and Drosophila.

10. Mutations are not limited in their time of occurrence to any particular period of the life history. This has been proved in the above-mentioned studies on mutable plants, in Drosophila, and in other cases.

11. Genes normal to the species tend to have more dominance than the mutant genes arising from them. This is very markedly the case in Drosophila, where even the relatively few mutant genes that have been called dominant are very incompletely so, and might more justly be called recessive. In other organisms, the same condition of things is strongly suggested, although the direct data on occurrence of mutations is as yet too meagre to allow of certainty.

12. Most mutations are deleterious in their effects. This applies not only to the organism as a whole but also to the development of any particular part: the delicate mechanisms for producing characters are more likely to be upset than strengthened, so that mutations should more often result in apparent losses or retrogressions than in "progressive" changes. This is both an a priori expectation and a phenomenon generally observed.

13. Mutations with slight effects are probably more frequent than those with more marked effects. This must not be understood as referring to the different mutations of each given gene, but it applies in a comparison of the mutations occurring in different genes. Thus, there are more than a dozen mutations, in different loci, which reduce the size of the wing in Drosophila so slightly as to leave it more than half its original length, whereas only four reduce it to less than half-length. Mutant genes with effects so slight as to be visible only by the aid of specific co-genes seem to arise still more frequently. It is reasonable to conclude that the mutations with slighter effects would more often take part in evolution, because they should usually be less deleterious, and this conclusion is borne out by observations on the multi-

plicity with which such factor-differences with relatively slight effects are found in species crosses.

14. The range of those mutations which are of appropriate magnitude to be visible is probably very small, in comparison with the entire "spectrum" of mutations, so that there are many more lethals than visible mutations, and probably more subliminal than visible.

The above empirical and semi-empirical principles must be regarded as a mere preliminary scaffolding, for the erection of a later, more substantial, theory of mutation. Time does not permit me here to discuss which directions of research, and what methods, seem the most promising for future results. Suffice it to say that it is especially important to obtain accurate data concerning the effect of various conditions upon the rate of mutation. This seems one of the logical routes by which to work towards the artificial production of mutation and consequent more perfect control of evolution. At the same time such results should also give a further insight into the structure of the gene. The way is now open, for the first time, to such studies on mutation rate, first through the finding, by Emerson, Baur, Maryatt, Zeleny, and Blakeslee of a number of specially mutable factors in different organisms, and second, through certain special genetic methods which I have elaborated in Drosophila, for the detection of lethal and other mutations there.

It has now become recognized that advances in theoretical or "pure" science eventually carry in their train changes in practice of the most far-reaching nature—changes which are usually far more radical than those caused by progress in the applied science directly concerned. It may therefore be asked at this point by eugenists: "Are there any applications of the knowledge which has already been gained about mutation in general, to eugenics and to the principles which should govern us in guiding human reproduction?" I think that one such application is already clearly indicated.

In order to understand the nature of this application it will be necessary first to consider the proposition—emphasized by East and Jones in their book, "Inbreeding and Crossbreeding"—that the only way for a genetically sound stock to be formed is by its going through a course of inbreeding, with elimination, by natural or artificial selection, of the undesirable individuals that appear in the course of this inbreeding. The truth of this propo-

sition depends upon the fact that many recessive genes of undesirable character are apt to exist in a population. Since the frequency with which these genes are able to produce their characteristic effects, i.e., to "come to light," depends on the closeness of the inbreeding, it is evident that inbreeding will be necessary in order to recognize the genes adequately, and hence to eliminate them.

Our present theory of mutation, however, carries us further than the proposition just considered. It shows that these undesirable genes have arisen by mutation; in fact, as stated in point 12, the great majority of mutations are deleterious, probable even to the degree of being lethal, and it is also known, as noted in point 11, that many—probably the great majority—are recessive. In other words, our mutation theory shows that probably the majority of the mutations that are occurring are giving rise to genes of just the type specified in the above discussion. This immediately shows us that not only are inbreeding and selection desirable for raising the genetic level of a population, but they are absolutely necessary merely in order to maintain it at its present standard. For the same process of mutation which was responsible for the origination of these undesirable genes in the past must be producing them now, and will continue to produce them in the future. Therefore, without selection, or without the inbreeding that makes effective selection possible, these lethals and other undesirable genes will inevitably accumulate, until the germ plasm becomes so riddled through with defect that pure lines cannot be obtained, and progress through selection of desirable recessive traits can never more be effected, since each of them will have become tied up with a lethal. To avoid such a complete and permanent collapse of the evolutionary process, it is accordingly necessary for man or nature to resort to a periodically repeated, although not continuous, series of inbreedings and selections in the case of any biparental organism.

This conclusion is more than a mere speculation, or even a deduction from our principles. The reality of this process of mutational deterioration has been directly proved, in the case of Drosophila, through experiments that I have conducted on lines in which the processes that usually accompany inbreeding and selection were prevented: in these lines there was found an accumulation of lethal genes so rapid that it would have taken but a few decades to have brought about the presence of a lethal

gene in practically every chromosome of every fly. Although the same general thesis undoubtedly applies also to mankind we do not yet know the speed of the process here. Its speed depends upon the actual frequency of mutations, which it will be very important—and extremely difficult—to determine in the case of mankind. Meanwhile, no matter what this rate may be, the process remains a real one, which must eventually be reckoned with, and either grappled in time, and conquered, or else yielded to.

I have dwelt at length upon this particular application to eugenics of some of the mutation studies. I believe, however, that this is but one example of such applications, and that from an increasing knowledge of our theoretical science there will inevitably flow an increasingly adequate technique for coping with our refractory human material. Meanwhile, the crying need is for more of the theoretical knowledge—and for the support of pure science, in its investigation of the processes lying at the root of the germ plasm.

1923

NOTE

[1] Contiguous genes may be affected in the rare cases known as "deficiences," found by Bridges and Mohr.

G. W. Beadle and E. L. Tatum

Genetic Control of Biochemical Reactions in Neurospora

On the basis of the characters expressed by mutations in flies and maize, it was reasonable to conclude that no organ system, no tissue, no stage of the life cycle, and no known physiological process was immune from genetic control. But the mechanism by which the genes controlled heredity was unknown. In 1903, Sir Archibald Garrod proposed a class of diseases which he called "inborn errors of metabolism." These were inherited abnormalities which caused, among other things, albinism and the inability to utilize certain amino acids in the body metabolism. Biochemistry, however, was not refined enough to work out the basis for any of these defects. During the 1920's and 1930's biochemists realized the importance of enzymes in synthesizing various small molecules such as sugars, vitamins, and amino acids.

George Beadle, a geneticist by training, took a strong interest in the biochemical basis of heredity. He first showed that in the fruit fly eye-color pigment is synthesized by several genes. With the association of Ephrussi and Tatum, Beadle identified some of these biochemical products. The fruit fly, however, was not a suitable organism for biochemical studies, and Beadle chose the fungus Neurospora. Using X rays, Beadle attempted to repeat Muller's induction of mutations. Beadle did not choose any mutation that chance provided; he selected those mutations that affected the synthesis of specific vitamins. The fact that mutations did occur and that these did affect the synthesis of only one small molecule, led Beadle and Tatum to an interesting theory. They proposed a "one gene–one enzyme" theory. Each gene affected one step in a biochemical pathway. Presumably this meant that the mutant gene controlled or produced a defective enzyme which could not carry out its metabolic function. Beadle and Tatum

Reprinted from *Proceedings of the National Academy of Sciences* 27:499–506, 1941, by permission of the Academy and the authors.

were able to extend their technique to other molecules. Mutations affected amino acids, fats, purines, carbohydrates, and all other physiological components of the cell. The influence of Beadle's and Tatum's results was widespread. Bacteria, viruses, and algae could now be treated as genetic organisms. Microbial genetics became the chief interest of biochemists hoping to analyze the nature of the gene.

From the standpoint of physiological genetics the development and functioning of an organism consist essentially of an integrated system of chemical reactions controlled in some manner by genes. It is entirely tenable to suppose that these genes which are themselves a part of the system, control or regulate specific reactions in the system either by acting directly as enzymes or by determining the specificities of enzymes.[1] Since the components of such a system are likely to be interrelated in complex ways, and since the synthesis of the parts of individual genes are presumably dependent on the functioning of other genes, it would appear that there must exist orders of directness of gene control ranging from simple one-to-one relations to relations of great complexity. In investigating the rôles of genes, the physiological geneticist usually attempts to determine the physiological and biochemical bases of already known hereditary traits. This approach, as made in the study of anthocyanin pigments in plants,[2] the fermentation of sugars by yeasts[3] and a number of other instances,[4] has established that many biochemical reactions are in fact controlled in specific ways by specific genes. Furthermore, investigations of this type tend to support the assumption that gene and enzyme specificities are of the same order.[5] There are, however, a number of limitations inherent in this approach. Perhaps the most serious of these is that the investigator must in general confine himself to a study of non-lethal heritable characters. Such characters are likely to involve more or less non-essential so-called "terminal" reactions.[5] The selection of these for genetic study was perhaps responsible for the now rapidly disappearing belief that genes are concerned only with the control of "superficial" characters. A second difficulty, not unrelated to the first, is that the standard approach to the problem implies the use of characters with visible manifestations. Many such characters involve morphological variations, and these are likely to be based on systems of biochemical reactions so complex as to make analysis exceedingly difficult.

Considerations such as those just outlined have led us to investi-

gate the general problem of the genetic control of developmental and metabolic reactions by reversing the ordinary procedure and, instead of attempting to work out the chemical bases of known genetic characters, to set out to determine if and how genes control known biochemical reactions. The ascomycete *Neurospora* offers many advantages for such an approach and is well suited to genetic studies.[6] Accordingly, our program has been built around this organism. The procedure is based on the assumption that X-ray treatment will induce mutations in genes concerned with the control of known specific chemical reactions. If the organism must be able to carry out a certain chemical reaction to survive on a given medium, a mutant unable to do this will obviously be lethal on this medium. Such a mutant can be maintained and studied, however, if it will grow on a medium to which has been added the essential product of the genetically blocked reaction. The experimental procedure based on this reasoning can best be illustrated by considering a hypothetical example. Normal strains of *Neurospora crassa* are able to use sucrose as a carbon source, and are therefore able to carry out the specific and enzymatically controlled reaction involved in the hydrolysis of this sugar.

Assuming this reaction to be genetically controlled, it should be possible to induce a gene to mutate to a condition such that the organism could no longer carry out sucrose hydrolysis. A strain carrying this mutant would then be unable to grow on a medium containing sucrose as a sole carbon source but should be able to grow on a medium containing some other normally utilizable carbon source. In other words, it should be possible to establish and maintain such a mutant strain on a medium containing glucose and detect its inability to utilize sucrose by transferring it to a sucrose medium.

Essentially similar procedures can be developed for a great many metabolic processes. For example, ability to synthesize growth factors (vitamins), amino acids and other essential substances should be lost through gene mutation if our assumptions are correct. Theoretically, any such metabolic deficiency can be "by-passed" if the substance lacking can be supplied in the medium and can pass cell walls and protoplasmic membranes.

In terms of specific experimental practice, we have devised a procedure in which X-rayed single-spore cultures are established on a so-called "complete" medium, i.e., one containing as many

of the normally synthesized constituents of the organism as is practicable. Subsequently these are tested by transferring them to a "minimal" medium, i.e., one requiring the organism to carry on all the essential syntheses of which it is capable. In practice the complete medium is made up of agar, inorganic salts, malt extract, yeast extract and glucose. The minimal medium contains agar (optional), inorganic salts and biotin, and a disaccharide, fat or more complex carbon source. Biotin, the one growth factor that wild type *Neurospora* strains cannot synthesize,[7] is supplied in the form of a commercial concentrate containing 100 micrograms of biotin per cc.[8] Any loss of ability to synthesize an essential substance present in the complete medium and absent in the minimal medium is indicated by a strain growing on the first and failing to grow on the second medium. Such strains are then tested in a systematic manner to determine what substance or substances they are unable to synthesize. These subsequent tests include attempts to grow mutant strains on the minimal medium with (1) known vitamins added, (2) amino acids added or (3) glucose substituted for the more complex carbon source of the minimal medium.

Single ascospore strains are individually derived from perithecia of *N. crassa* and *N. sitophila* X-rayed prior to meiosis. Among approximately 2000 such strains, three mutants have been found that grow essentially normally on the complete medium and scarcely at all on the minimal medium with sucrose as the carbon source. One of these strains (*N. sitophila*) proved to be unable to synthesize vitamin B_6 (pyridoxine). A second strain (*N. sitophila*) turned out to be unable to synthesize vitamin B_1 (thiamine). Additional tests show that this strain is able to synthesize the pyrimidine half of the B_1 molecule but not the thiazole half. If thiazole alone is added to the minimal medium, the strain grows essentially normally. A third strain (*N. crassa*) has been found to be unable to synthesize para-aminobenzoic acid. This mutant strain appears to be entirely normal when grown on the minimal medium to which p-aminobenzoic acid has been added. Only in the case of the "pyridoxinless" strain has an analysis of the inheritance of the induced metabolic defect been investigated. For this reason detailed accounts of the thiamine-deficient and p-aminobenzoic acid–deficient strains will be deferred.

Qualitative studies indicate clearly that the pyridoxinless mutant, grown on a medium containing one microgram or more of synthetic vitamin B_6 hydrochloride per 25 cc. of medium, closely

approaches in rate and characteristics of growth normal strains grown on a similar medium with no B_6. Lower concentrations of B_6 give intermediate growth rates. A preliminary investigation of the quantitative dependence of growth of the mutant on vitamin B_6 in the medium gave the results summarized in Table I.

TABLE I

GROWTH OF PYRIDOXINLESS STRAIN OF *N. sitophila* ON LIQUID MEDIUM CONTAINING INORGANIC SALTS,[9] 1% SUCROSE, AND 0.004 MICROGRAM BIOTIN PER Cc. TEMPERATURE 25°C. GROWTH PERIOD, 6 DAYS FROM INOCULATION WITH CONIDIA

Micrograms B_6 per 25 cc. Medium	Strain	Dry Weight Mycelia, mg.
0	Normal	76.7
0	Pyridoxinless	1.0
0.01	"	4.2
0.03	"	5.7
0.1	"	13.7
0.3	"	25.5
1.0	"	81.1
3.0	"	81.1
10.0	"	65.4
30.0	"	82.4

Additional experiments have given results essentially similar but in only approximate quantitative agreement with those of Table I. It is clear that additional study of the details of culture conditions is necessary before rate of weight increase of this mutant can be used as an accurate assay for vitamin B_6.

It has been found that the progression of the frontier of mycelia of *Neurospora* along a horizontal glass culture tube half filled with agar medium provides a convenient method of investigating the quantitative effects of growth factors. Tubes of about 13 mm. inside diameter and about 40 cm. in length are used. Segments of about 5 cm. at the two ends are turned up at an angle of about 45°. Agar medium is poured in so as to fill the tube about half

full and is allowed to set with the main segment of the tube in a horizontal position. The turned up ends of the tube are stoppered with cotton plugs. Inoculations are made at one end of the agar surface and the position of the advancing front recorded at convenient intervals. The frontier formed by the advancing mycelia is remarkably well defined, and there is no difficulty in determining its position to within a millimeter or less. Progression along such tubes is strictly linear with time and the rate is independent of tube length (up to 1.5 meters). The rate is not changed by reducing the inside tube diameter to 9 mm., or by sealing one or both ends. It therefore appears that gas diffusion is in no way limiting in such tubes.

The results of growing the pyridoxinless strain in horizontal tubes in which the agar medium contained varying amounts of B_6 are shown graphically in Figures 1 and 2. Rate of progression is clearly a function of vitamin B_6 concentration in the medium.[10] It is likewise evident that there is no significant difference in rate between the mutant supplied with B_6 and the normal strain growing on a medium without this vitamin. These results are consistent with the assumption that the primary physiological difference between pyridoxinless and normal strains is the inability of the former to carry out the synthesis of vitamin B_6. There is certainly more than one step in this synthesis and accordingly the gene differential involved is presumably concerned with only one specific step in the biosynthesis of vitamin B_6.

In order to ascertain the inheritance of the pyridoxinless character, crosses between normal and mutant strains were made. The techniques for hybridization and ascospore isolation have been worked out and described by Dodge, and by Lindegren.[6] The ascospores fom 24 asci of the cross were isolated and their positions in the asci recorded. For some unknown reason, most of these failed to germinate. From seven asci, one or more spores germinated. These were grown on a medium containing glucose, malt extract and yeast extract, and in this they all grew normally. The normal and mutant cultures were differentiated by growing them on a B_6-deficient medium. On this medium the mutant cultures grew very little, while the non-mutant ones grew normally. The results are summarized in Table II. It is clear from these rather limited data that this inability to synthesize vitamin B_6 is transmitted as it should be if it were differentiated from normal by a single gene.

The preliminary results summarized above appear to us to indicate the approach outlined may offer considerable promise as a method of learning more about how genes regulate development and function. For example, it should be possible, by finding a number of mutants unable to carry out a particular step in a given synthesis, to determine whether only one gene is ordinarily concerned with the immediate regulation of a given specific chemical reaction.

TABLE II

RESULTS OF CLASSIFYING SINGLE ASCOSPORE CULTURES FROM THE CROSS OF PYRIDOXINLESS AND NORMAL *N. sitophila*

ASCUS NUMBER	1	2	3	4	5	6	7	8
17	—	pdx	pdx	pdx	N	N	N	—
18	—	—	N	N	—	—	pdx	pdx
19	—	pdx	—	—	—	—	—	N
20	—	—	N	—	—	—	—	pdx
22	—	—	N	—	—	—	—	—
23	—	*	*	*	N	N	pdx	pdx
24	N	N	N	N	pdx	pdx	pdx	pdx

N, normal growth on B_6-free medium. pdx, slight growth on B_6-free medium. Failure of ascospore germination indicated by dash.

* Spores 2, 3 and 4 isolated but positions confused. Of these, two germinated and both proved to be mutants.

It is evident, from the standpoints of biochemistry and physiology, that the method outlined is of value as a technique for discovering additional substances of physiological significance. Since the complete medium used can be made up with yeast extract or with an extract of normal *Neurospora,* it is evident that if, through mutation, there is lost the ability to synthesize an essential substance, a test strain is thereby made available for use in isolating the substance. It may, of course, be a substance not previously known to be essential for the growth of any organism. Thus we may expect to discover new vitamins, and in the same way, it should be possible to discover additional essential

amino acids if such exist. We have, in fact, found a mutant strain that is able to grow on a medium containing Difco yeast extract but unable to grow on any of the synthetic media we have so far tested. Evidently some growth factor present in yeast and as yet unknown to us is essential for *Neurospora*.

Summary.—A procedure is outlined by which, using *Neurospora,* one can discover and maintain X-ray induced mutant strains which are characterized by their inability to carry out specific biochemical processes.

Following this method, three mutant strains have been established. In one of these the ability to synthesize vitamin B_6 has been wholly or largely lost. In a second the ability to synthesize the thiazole half of the vitamin B_1 molecule is absent, and in the third para-aminobenzoic acid is not synthesized. It is therefore clear that all of these substances are essential growth factors for *Neurospora.*[11]

Growth of the pyridoxinless mutant (a mutant unable to synthesize vitamin B_6) is a function of the B_6 content of the medium on which it is grown. A method is described for measuring the growth by following linear progression of the mycelia along a horizontal tube half filled with an agar medium.

Inability to synthesize vitamin B_6 is apparently differentiated by a single gene from the ability of the organism to elaborate this essential growth substance.

NOTE: Since the manuscript of this paper was sent to press it has been established that inability to synthesize both thiazole and p-aminobenzoic acid is also inherited as though differentiated from normal by single genes.

This work was supported in part by a grant from the Rockefeller Foundation. The authors are indebted to Doctors B. O. Dodge, C. C. Lindegren and W. S. Malloch for stocks and for advice on techniques, and to Miss Caryl Parker for technical assistance.

1941

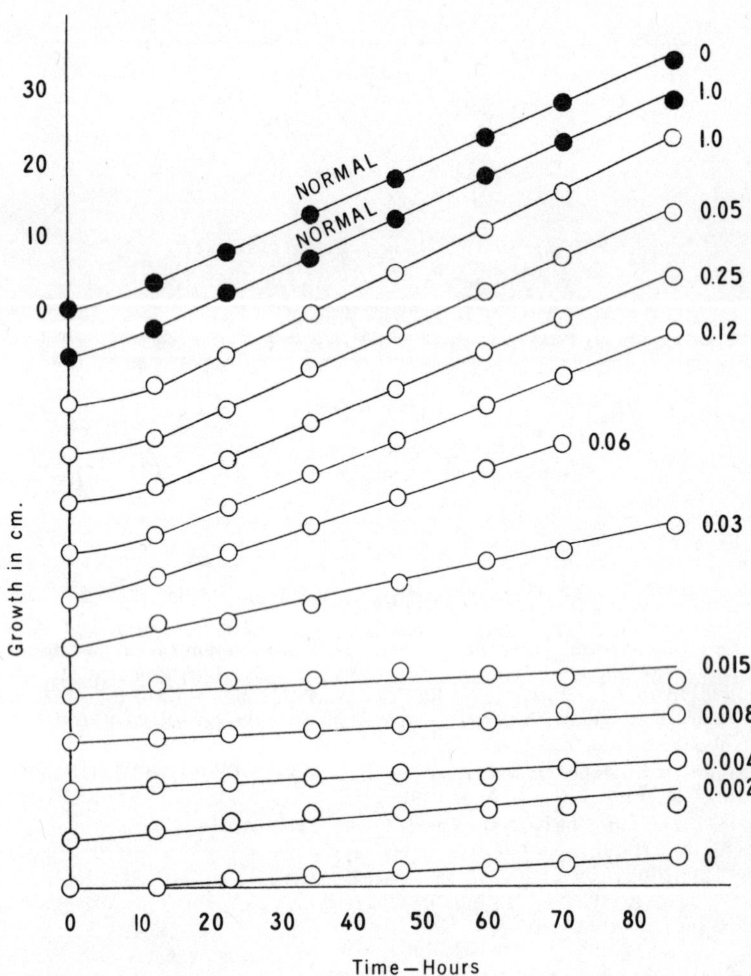

FIG. 1. Growth of normal (top two curves) and pyridoxinless (remaining curves) strains of *Neurospora sitophila* in horizontal tubes. The scale on the ordinate is shifted a fixed amount for each successive curve in the series. The figures at the right of each curve indicate concentration of pyridoxin (B_6) in micrograms per 25 cc. medium.

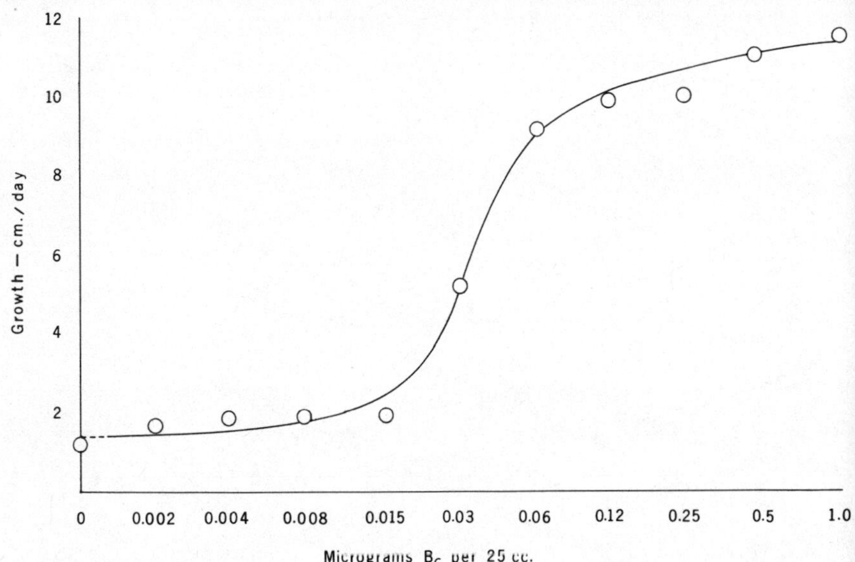

FIG. 2. The relation between growth rate (cm./day) and Vitamin B_6 concentration.

NOTES

1 The possibility that genes may act through the mediation of enzymes has been suggested by several authors. See Troland, L. T., *Amer. Nat.*, 51, 321–350 (1917); Wright, S., *Genetics*, 12, 530–569 (1927); and Haldane, J. B. S., in *Perspectives in Biochemistry*, Cambridge Univ. Press, pp. 1–10 (1937), for discussions and references.

2 Onslow, Scott-Moncrieff and others; see review by Lawrence, W. J. C., and Price, J. R., *Biol. Rev.*, 15, 35–58 (1940).

3 Winge, O., and Laustsen, O., *Compt. rend. Lab. Carlsberg, Serie physiol.*, 22, 337–352 (1939).

4 See Goldschmidt, R., *Physiological Genetics*, McGraw-Hill, pp. 1–375 (1939), and Beadle, G. W., and Tatum, E. L., *Amer. Nat.*, 75, 107–116 (1941) for discussion and references.

5 See Sturtevant, A. H., and Beadle, G. W., *An Introduction to Genetics*, Saunders, pp. 1–391 (1931), and Beadle, G. W., and Tatum, E. L., *loc. cit.*, footnote 4.

6 Dodge, B. O., *Journ. Agric. Res.*, 35, 289–305 (1927), and Lindegren, C. C., *Bull. Torrey Bot. Club*, 59, 85–102 (1932).

7 Insofar as we have carried them, our investigations on the vitamin requirements of *Neurospora* corroborate those of Butler, E. T., Robbins, W. J., and Dodge, B. O., *Science*, 94, 262–263 (1941).

[8] The biotin concentrate used was obtained from the S. M. A. Corporation, Chagrin Falls, Ohio.

[9] Throughout our work with *Neurospora,* we have used as a salt mixture the one designated number 3 by Fries, N., *Symbolae Bot. Upsalienses,* Vol. 3, No. 2, 1–188 (1938). This has the following composition: NH_4 tartrate, 5 g.; NH_4NO_3, 1 g; KH_2PO_4, 1 g.; $MgSO_4 \cdot 7H_2O$, 0.5 g.; NaCl, 0.1 g.; $CaCl_2$, 0.1 g.; $FeCl_3$, 10 drops 1% solution; H_2O, 1 L. The tartrate cannot be used as a carbon source by *Neurospora.*

[10] It is planned to investigate further the possibility of using the growth of *Neurospora* strains in the described tubes as a basis of vitamin assay, but it should be emphasized that such additional investigation is essential in order to determine the reproducibility and reliability of the method.

[11] The inference that the three vitamins mentioned are essential for the growth of normal strains is supported by the fact that an extract of the normal strain will serve as a source of vitamin for each of the mutant strains.

J. D. Watson and F. H. C. Crick

Genetical Implications of the Structure of Deoxyribonucleic Acid

In 1921 Muller emphasized the important, if not unique, feature which distinguished genes from all other compounds. This was "convariant reproduction," the ability of the gene to retain its capacity for reproduction despite some alteration leading to a changed function. In the early 1950's DNA was recognized as the chemical basis of heredity, but no one knew how DNA replicated, how it carried the biological specificity which distinguished each gene, nor how the process of mutation could be explained by alterations of DNA. J. D. Watson studied genetics at Indiana University where he absorbed the influence of Muller, Sonneborn, Cleland, and Luria, each a world authority in his special field of genetics. Watson found bacteriophage studies in Luria's laboratory especially rewarding. While on a postdoctoral fellowship, in Cambridge, England, he collaborated with F. H. C. Crick, a biophysicist. They used X-ray crystallography to supplement their theoretical inferences that the DNA molecule was a double molecule. The concept of complementary pairing of purine and pyrimidine bases provided the mechanism for replication. The linear sequence of base pairs provided the basis of a genetic code and the specificity of the gene. Mutation was recognized to be no more complex than a substitution or change of one base pair for another. In this second paper, published in 1953, Watson and Crick discuss the biological significance of the model they proposed for the structure of DNA. Within five years after publication of the Watson-Crick model the implications of its mode of replication and the mechanism of mutation through pairing errors were verified. By 1962 the details of the genetic code were worked out, and once again the Watson-Crick model provided the basis for its interpretation. No single event in the history of biology

Reprinted from *Nature 171*:964, 1953, by permission of the publisher and the authors.

has done more to bring together biologists, physicists, and chemists, than this model. The field of molecular biology rapidly developed after the publication of the Watson-Crick model. The award of the Nobel Prize in physiology was made in 1963 for this remarkable contribution.

The importance of deoxyribonucleic acid (DNA) within living cells is undisputed. It is found in all dividing cells, largely if not entirely in the nucleus, where it is an essential constituent of the chromosomes. Many lines of evidence indicate that it is the carrier of a part of (if not all) the genetic specificity of the chromosomes and thus of the gene itself. Until now, however, no evidence has been presented to show how it might carry out the essential operation required of a genetic material, that of exact self-duplication.

We have recently proposed a structure[1] for the salt of deoxyribonucleic acid, which, if correct, immediately suggests a mechanism for its self-duplication. X-ray evidence obtained by the workers at King's College, London,[2] and presented at the same time, gives qualitative support to our structure and is incompatible with all previously proposed structures.[3] Though the structure will not be completely proved until a more extensive comparison has been made with the X-ray data, we now feel sufficient confidence in its general correctness to discuss its genetical implications. In doing so we are assuming that fibres of the salt of deoxyribonucleic acid are not artefacts arising in the method of preparation, since it has been shown by Wilkins and his co-workers that similar X-ray patterns are obtained from both the isolated fibres and certain intact biological materials such as sperm head and bacteriophage particles.[2, 4]

The chemical formula of deoxyribonucleic acid is now well established. The molecule is a very long chain, the backbone of which consists of a regular alternation of sugar and phosphate groups, as shown in Fig. 1. To each sugar is attached a nitrogenous base, which can be of four different types. (We have considered 5-methyl cytosine to be equivalent to cytosine, since either can fit equally well into our structure.) Two of the possible bases—adenine and guanine—are purines, and the other two—thymine and cytosine—are pyrimidines. So far as is known, the sequence of bases along the chain is irregular. The monomer unit, consisting of phosphate, sugar and base, is known as a nucleotide.

The first feature of our structure which is of biological interest is that it consists not of one chain, but of two. These two chains are both coiled around a common fibre axis, as is shown diagrammatically in Fig. 2. It has often been assumed that since there was only one chain in the chemical formula there would only be one in the structural unit. However, the density, taken with the X-ray evidence,[2] suggests very strongly that there are two.

The other biologically important feature is the manner in which the two chains are held together. This is done by hydrogen bonds between the bases, as shown schematically in Fig. 3. The bases are joined together in pairs, a single base from one chain being hydrogen-bonded to a single base from the other. The important point is that only certain pairs of bases will fit into the structure One member of a pair must be a purine and the other a pyrimidine in order to bridge between the two chains. If a pair consisted of two purines, for example, there would not be room for it.

We believe that the bases will be present almost entirely in their most probable tautomeric forms. If this is true, the conditions for forming hydrogen bonds are more restrictive, and the only pairs of bases possible are:

adenine with thymine;
guanine with cytosine.

The way in which these are joined together is shown in Figs. 4 and 5. A given pair can be either way round. Adenine, for example, can occur on either chain; but when it does, its partner on the other chain must always be thymine.

This pairing is strongly supported by the recent analytical results,[5] which show that for all sources of deoxyribonucleic acid examined the amount of adenine is close to the amount of thymine, and the amount of guanine close to the amount of cytosine, although the cross-ratio (the ratio of adenine to guanine) can vary from one source to another. Indeed, if the sequence of bases on one chain is irregular, it is difficult to explain these analytical results except by the sort of pairing we have suggested.

The phosphate-sugar backbone of our model is completely regular, but any sequence of the pairs of bases can fit into the structure. It follows that in a long molecule many different permutations are possible, and it therefore seems likely that the precise sequence of the bases is the code which carries the genetical information. If the actual order of the bases on one of the pair of chains were given, one could write down the exact order of

the bases on the other one, because of the specific pairing. Thus one chain is, as it were, the complement of the other, and it is this feature which suggests how the deoxyribonucleic acid molecule might duplicate itself.

Previous discussions of self-duplication have usually involved the concept of a template, or mould. Either the template was supposed to copy itself directly or it was to produce a 'negative', which in its turn was to act as a template and produce the original 'positive' once again. In no case has it been explained in detail how it would do this in terms of atoms and molecules.

Now our model for deoxyribonucleic acid is, in effect, a pair of templates, each of which is complementary to the other. We imagine that prior to duplication the hydrogen bonds are broken, and the two chains unwind and separate. Each chain then acts as a template for the formation on to itself of a new companion chain, so that eventually we shall have two pairs of chains, where we only had one before. Moreover, the sequence of the pairs of bases will have been duplicated exactly.

A study of our model suggests that this duplication could be done most simply if the single chain (or the relevant portion of it) takes up the helical configuration. We imagine that at this stage in the life of the cell, free nucleotides, strictly polynucleotide precursors, are available in quantity. From time to time the base of a free nucleotide will join up by hydrogen bonds to one of the bases on the chain already formed. We now postulate that the polymerization of these monomers to form a new chain is only possible if the resulting chain can form the proposed structure. This is plausible, because steric reasons would not allow nucleotides 'crystallized' on to the first chain to approach one another in such a way that they could be joined together into a new chain, unless they were those nucleotides which were necessary to form our structure. Whether a special enzyme is required to carry out the polymerization, or whether the single helical chain already formed acts effectively as an enzyme, remains to be seen.

Since the two chains in our model are intertwined, it is essential for them to untwist if they are to separate. As they make one complete turn around each other in 34 Å., there will be about 150 turns per million molecular weight, so that whatever the precise structure of the chromosome a considerable amount of uncoiling would be necessary. It is well known from microscopic observation that much coiling and uncoiling occurs during mitosis, and though this is on a much larger scale it probably reflects

similar processes on a molecular level. Although it is difficult at the moment to see how these processes occur without everything getting tangled, we do not feel that this objection will be insuperable.

Our structure, as described,[1] is an open one. There is room between the pair of polynucleotide chains (see Fig. 2) for a polypeptide chain to wind around the same helical axis. It may be significant that the distance between adjacent phosphorus atoms, 7·1 Å., is close to the repeat of a fully extended polypeptide chain. We think it probable that in the sperm head, and in artificial nucleoproteins, the polypeptide chain occupies this position. The relative weakness of the second layer-line in the published X-ray pictures [3a,1] is crudely compatible with such an idea. The function of the protein might well be to control the coiling and uncoiling, to assist in holding a single polynucleotide chain in a helical configuration, or some other non-specific function.

Our model suggests possible explanations for a number of other phenomena. For example, spontaneous mutation may be due to a base occasionally occurring in one of its less likely tautomeric forms. Again, the pairing between homologous chromosomes at meiosis may depend on pairing between specific bases. We shall discuss these ideas in detail elsewhere.

For the moment, the general scheme we have proposed for the reproduction of deoxyribonucleic acid must be regarded as speculative. Even if it is correct, it is clear from what we have said that much remains to be discovered before the picture of genetic duplication can be described in detail. What are the polynucleotide precursors? What makes the pair of chains unwind and separate? What is the precise role of the protein? Is the chromosome one long pair of deoxyribonucleic acid chains, or does it consist of patches of the acid joined together by protein?

Despite these uncertainties we feel that our proposed structure for deoxyribonucleic acid may help to solve one of the fundamental biological problems—the molecular basis of the template needed for genetic replication. The hypothesis we are suggesting is that the template is the pattern of bases formed by the chain of deoxyribonucleic acid and that the gene contains a complementary pair of such templates.

One of us (J.D.W.) has been aided by a fellowship from the National Foundation for Infantile Paralysis (U.S.A.).

1953

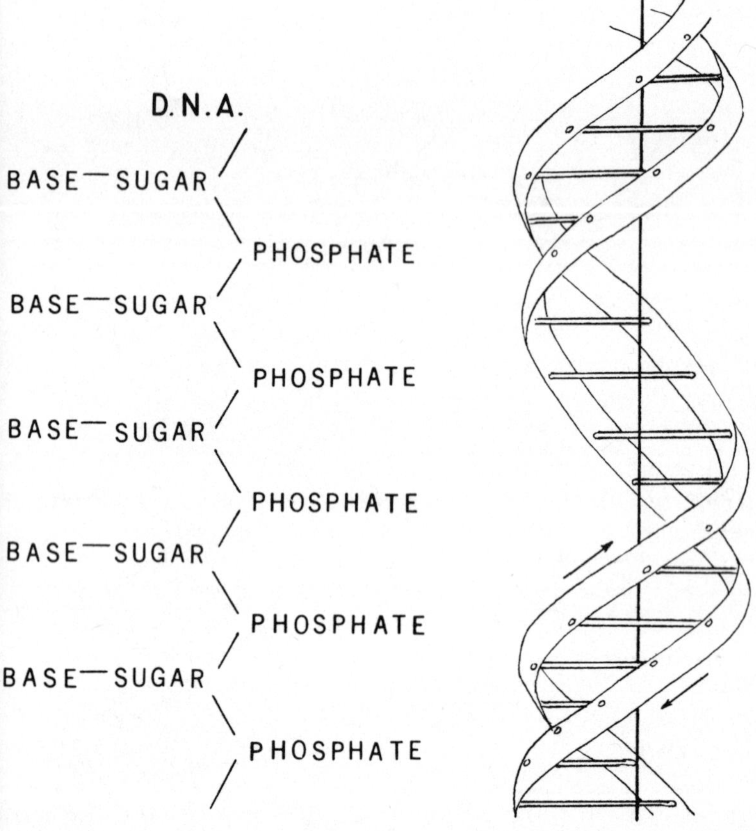

FIG. 1. Chemical formula of a single chain of deoxyribonucleic acid.

FIG. 2. This figure is purely diagrammatic. The two ribbons symbolize the two phosphate-sugar chains, and the horizontal rods the pairs of bases holding the chains together. The vertical line marks the fibre axis.

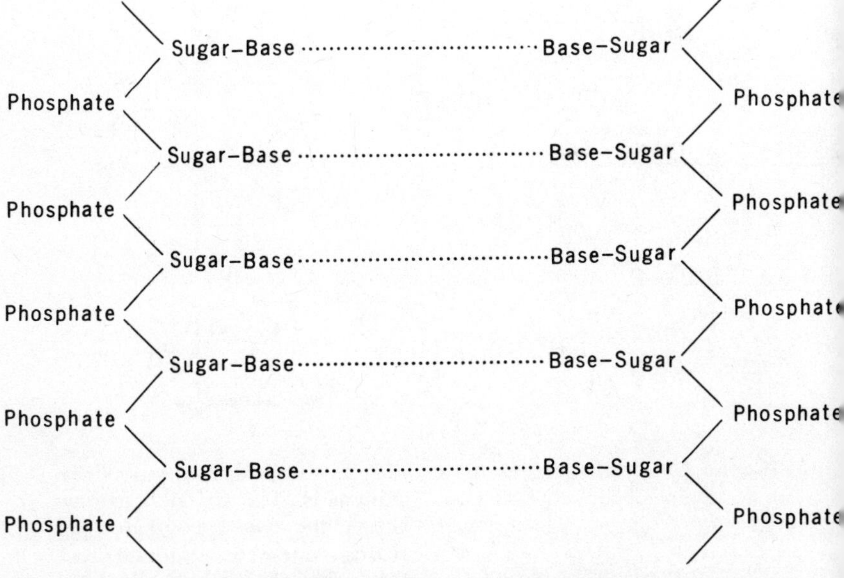

FIG. 3. Chemical formula of a pair of deoxyribonucleic acid chains. The hydrogen bonding is symbolized by dotted lines.

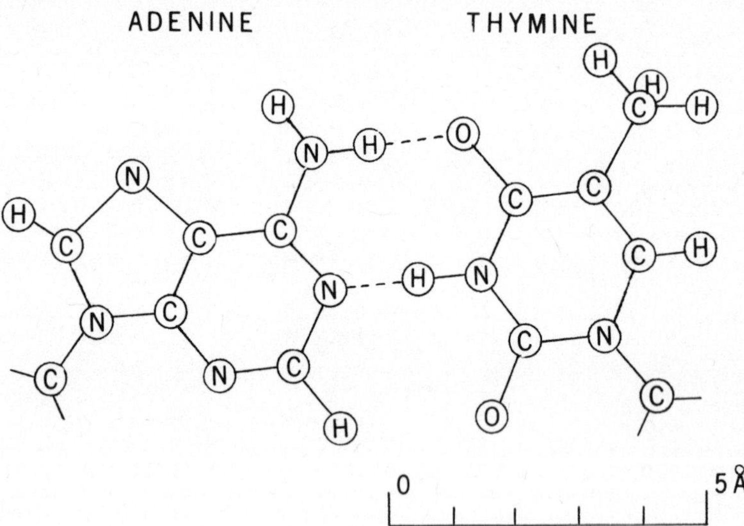

FIG. 4. Pairing of adenine and thymine. Hydrogen bonds are shown dotted. One carbon atom of each sugar is shown.

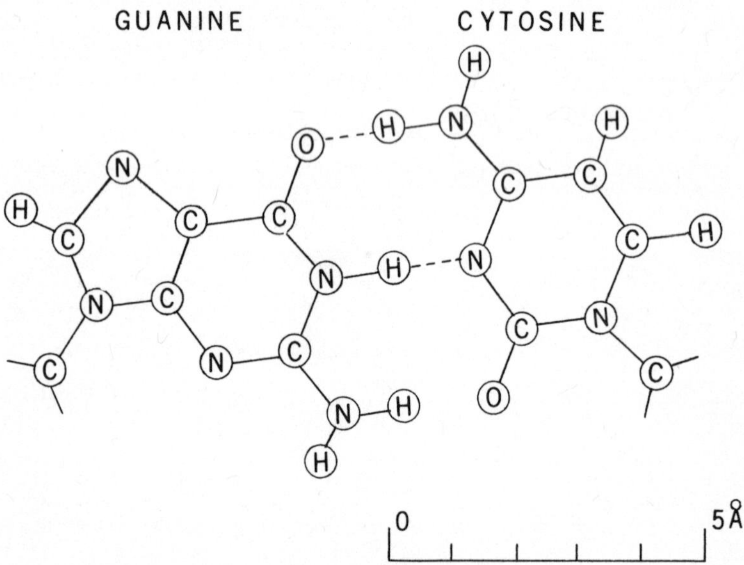

FIG. 5. Pairing of guanine and cytosine. Hydrogen bonds are shown dotted. One carbon atom of each sugar is shown.

NOTES

[1] Watson, J. D., and Crick, F. H. C., *Nature,* 171, 737 (1953).

[2] Wilkins, M. H. F., Stokes A. R., and Wilson, H. R., *Nature,* 171, 738 (1953). Franklin, R. E., and Gosling, R. G., *Nature,* 171, 740 (1953).

[3] (a) Astbury, W. T., *Symp. No. 1 Soc. Exp. Biol.,* 66 (1947). (b) Furberg, S., *Acta Chem. Scand.,* 6, 634 (1952). (c) Pauling, L., and Corey, R. B., *Nature,* 171, 346 (1953); *Proc. U.S. Nat. Acad. Sci.,* 39, 84 (1953). (d) Fraser, R. D. B. (in preparation).

[4] Wilkins, M. H. F., and Randall, J. T., *Biochim. et Biophys. Acta,* 10, 192 (1953).

[5] Chargaff, E.; for references see Zamenhof, S., Brawerman, G., and Chargaff, E., *Biochim. et Biophys. Acta,* 9, 402 (1952). Wyatt, G. R., *J. Gen. Physiol.,* 36, 201 (1952).

Seymour Benzer

Genetic Fine Structure and Its Relation to the DNA Molecule

Physicists have made numerous contributions to genetics. Mendel was a physicist by training and he taught physics in high school. Max Delbrück left physics to work on the molecular basis of life. He chose the bacteriophage virus and worked out much of its life cycle and the tools for making it a reliable laboratory organism. Seymour Benzer, a physicist, read another physicist's book, *What Is Life?* In it, Erwin Schrödinger discussed the work of Delbrück on mutation and gene stability. The stimulation led Benzer into biophysics and he learned bacteriophage techniques during a summer's residence at Cold Spring Harbor, New York. This background in physics and curiosity about gene structure gave Benzer an opportunity to work out an experiment in bacteriophage which resulted in a dramatic discovery. In 1955 Benzer reported the existence of "genetic fine structure." By using crossover data in a manner similar to that employed by Sturtevant some forty years before, Benzer found that the "gene" was divisible into hundreds of sites. There was also a limit to the smallest recombination frequency, and this value defined the limit of resolution of genetic analysis by crossing over. Benzer designated the gene as a physical unit. The functional gene he called the *cistron,* because it was defined genetically by a "cis-trans" test. The cistron is today used synonymously with the term *gene.* It has the connotation, however, that the gene has a fine structure. The fine structure of the virus gene corresponds to the nucleotide of the Watson-Crick DNA molecule. Benzer's discovery of genetic fine structure provided geneticists with a tool for the study of mutation and genetic coding. The remarkable feature of Benzer's work is that it uses genetic techniques throughout, in contrast

Reprinted from *Brookhaven Symposia in Biology* 8:3–5 (Report of Symposium on "Mutation," held June 15 to 17, 1955), by permission of the publisher and the author.

to many later developments in genetics which use molecular biology as the conceptual basis for formulating problems and biophysical and chemical tools for their experimental solution.

A signal achievement of genetics has been the demonstration that hereditary factors, the classical "genes," are ordered in a one-dimensional array in chromosomes divisible by genetic recombination. It is fascinating to inquire whether this one-dimensionality and divisibility persist down to the ultimate molecular structure of the genetic material. In most organisms, this inquiry is frustrated by a practical limitation on the attainable resolution of genetic recombination experiments—adequate resolution requires the detection of vanishingly small proportions of recombinant types. However, in bacterial viruses, the properties of certain mutants are sufficiently favorable to permit the achievement of this goal. The results so far obtained give encouraging support to the idea that a one-dimensional scheme can be extended even to the finest details of the genetic material. Furthermore, it seems feasible to arrive at clear characterizations of the ultimate units of physiological function, mutation, and genetic recombination, and to place an estimate upon the molecular size of each.

In another publication,[1] a description is given of the properties of the so-called rII group of mutants of bacteriophage T4 and the manner of performing experiments with them. Therefore, only a brief abstract will be given here.

Wild-type T4 produces similar plaques on two bacterial host strains of *Escherichia coli,* B and K. Mutants of the rII type are characterized by markedly altered plaques on B and fail to produce plaques on K. The mutant particles absorb to and kill K but do not complete the normal life cycle. All mutations of this type are found to be located within a small portion of the T4 linkage map (Figure 1A). Within that region, their locations, as determined by the frequency of production of wild recombinants in crosses between pairs of mutants, are widely scattered. An unambiguous seriation can be accomplished, except for certain anomalous cases. Mixed infection of a cell of strain K with a mutant and a wild-type particle leads to lysis with liberation of progeny of both types. Therefore, the rII mutation may be considered as recessive. A diploid double heterozygote in the trans configuration is simulated by infection of K with two different mutants. The application of the phenotype test[2] to pairs of rII

mutants leads to the division of the region into two functionally separable segments, as indicated in Figure 1B.

Selection of clusters-within-clusters for closer examination leads to the results exemplified by Figure 1, C and D. Two groups of mutants are shown which fail to give as much as 10^{-3} percent recombination with each other. Non-zero intervals as small as 10^{-2} percent have been found between mutant groups.

By comparing the total linkage length and total DNA content of a T4 particle, a very rough conversion factor may be calculated for translating linkage distances into molecular ones. This ratio is of the order of 10^{-3} percent recombination per nucleotide pair. According to this value, the functional segments of the rII region are about 4000 nucleotide pairs long. From the smallest non-zero distance between mutations, the chromosome would seem to be divisible by recombination at least to the extent of pieces ten nucleotide pairs in length. The size of a mutation, i.e., the length of chromosome in which alteration is apparent, varies from several hundred nucleotide pairs for the "anomalous" mutants to no more than ten for others.

Each rII mutant has a characteristic rate of reversion to wild type, the rates covering an enormous range (10^7-fold). The anomalous mutants are extremely stable. The possibility of suppressor mutations distant (i.e., more than around 100 nucleotide pairs) from the rII mutations has been ruled out by back crosses to the original wild type. Partial reversions to intermediate types are also observed. To widely varied degrees rII mutants are "leaky" (i.e., can multiply feebly on K) and some are temperature dependent.

These observations on a series of phage mutants are highly analogous to those seen in systems of "pseudoalleles" which have been found in many organisms. Demerec and Demerec, in this symposium, discuss a very comparable case in bacteria, and Giles one in *Neurospora*. It may not be unreasonable to hope that the structure and function of the genetic material in phage, which is so accessible to detailed study, can serve as a model for higher forms.

1955

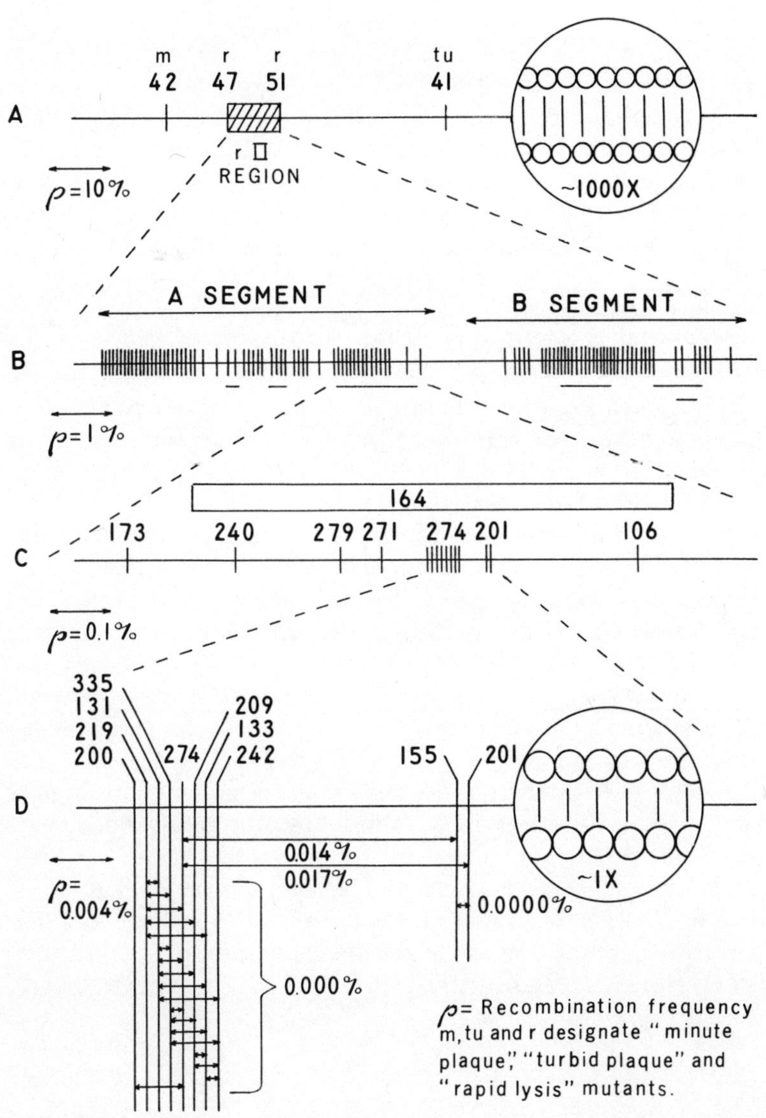

m
42
r
47
r
51
tu
41
A
r II
REGION
ρ=10%
~1000X
A SEGMENT
B SEGMENT
B
ρ=1%
164
173
240
279 271
274 201
106
C
ρ=0.1%
335
131
219
200
274
209
133
242
155
201
D
ρ=0.004%
0.014%
0.017%
0.0000%
0.000%
~1X
ρ= Recombination frequency
m, tu and r designate "minute plaque," "turbid plaque" and "rapid lysis" mutants.

FIG. 1. Linkage maps of T4 phage. A) The location of the rII region is shown with respect to mutants (mapped by Doermann and Hill[3]). The inset shows the estimated corresponding spacing of nucleotides (circles) according to the structure of Watson and Crick,[4] on the assumption that the phage "chromosome" is such a fiber of DNA. B) The locations are shown of a number of rII mutants within the region. (There is some distortion due to the difficulty of drawing lines sufficiently close to one another.) All mutants within a segment are functionally related, whereas the A and B segments are functionally independent. A horizontal line represents an "anomalous" mutant which produces no wild recombinants in crosses with any mutant covered by its span. C) A selected group is shown on larger scale. Wild recombinants are given by r164 with r173 but not with any of the others. D) Further magnification. The group of eight mutants on the left all have similar reversion rates and give less than 10^{-3} percent recombination with each other in the indicated crosses. They are presumed to be recurrences of identical mutations. Similarly, r155 and r201 are probably recurrences.

NOTES

1 Benzer, S., *Proc. Natl. Acad. Sci.*, 41, 344 (1955).

2 Lewis, E. B., *Cold Spring Harbor Symposia Quant. Biol.*, 16, 159 (1951); Pontecorvo, G., *Advances in Enzymol.*, 13, 121 (1952).

3 Doermann, A. H., and Hill, M. B., *Genetics*, 38, 79 (1953).

4 Watson, J. D., and Crick, F. H. C., *Cold Spring Harbor Symposia Quant. Biol.*, 18, 123 (1953).

PART III ooo

Development

SHORTLY AFTER THE DEVELOPMENT of microscopy, biologists observed the structure and life cycles of some of the smaller insects and microbial life. Swammerdam, a Dutch biologist, observed the eclosion (emergence) of adult insects from pupa cases and erroneously interpreted this as the emergence of an adult from an egg. Other biologists, repeating Leeuwenhoek's observations, confirmed the presence of spermatozoa in semen. These two observations led to a three-way controversy about the nature of development. One group, citing Aristotle's observations of developing chick embryos, claimed that development is *epigenetic*. The organs and tissues, according to the theory of epigenetic development, become more complex as the embryo develops, and neither the egg nor the sperm contains these organs or tissues. A second group, stimulated by Swammerdam and Bonnet, believed that the *egg* contains a miniature adult which merely grows larger; all the organs and tissues are present in the egg. This theory of *preformation* was an awkward interpretation of development because it required the encapsulation of an indefinite number of future generations of individuals, one inside the other. The third group of seventeenth century biologists looked upon the *sperm* as the source of preformation. The egg, to these investigators, is merely a nutritive organ in which the sperm develops. Microscopists with imagination even drew pictures of a *homonculus* in human sperm. In the mid-eighteenth century Casper Wolff, carefully observing developing embryos microscopically, concluded that development is clearly epigenetic. This resolved the three-way debate and eliminated the two theories of preformation.

In the nineteenth century the role of the cell in development was emphasized; sperm and eggs were unambiguously associated with fertilization; and the cleavage of fertilized eggs was followed microscopically. Weismann's theory of the germ plasm generated the cytological search for a reduction in chromosome number by meiosis. The constancy of chromosome distribution was observed by numerous investigators who studied mitosis as the basis of cell division. Experimental embryology was stimulated by the efforts of W. Roux and H. Driesch, who reached opposite conclusions about the fate of the first cell divisions in development. Roux thought each cell is differentiated immediately after the first mitosis following fertilization (mosaic development), while Driesch believed that every cell produced by these early cell divisions can give rise to a complete organism (regulative development).

In the twentieth century much of the effort of embryologists was directed toward proving that both mosaic development and regulative development occur in any organism. Spemann's analysis favored the idea that the nuclei of higher organisms are not differentiated in the early cell divisions. This potential of generating a complete, normal, individual was extended to later stages of cell division by Briggs and King and other investigators. They transplanted nuclei from the cells of older embryos into fertilized eggs whose nuclei had been removed.

A considerable effort was made in the 1920's and 1930's to find a chemical basis for the "organizer" which was assumed to set the pattern for tissue differentiation and organization. Spemann had found that certain tissues when transferred from one embryo to another could induce the formation of secondary embryos near the insertion site of the host embryo. The change in cells resulting in tissues is called differentiation. That differentiation is a chemical process is generally accepted by embryologists, but the processes by which cell differentiation takes place are not known. Three approaches have been studied in the past fifteen years. First, immunological techniques can detect the presence of molecular components of a tissue or organ before histological changes can be observed microscopically. Second, the operon theory of Jacob and Monod provides a model of regulation which is analogous to the process of cell differentiation. If genes are switched on and off for biochemical reactions in bacteria, why can't genes be switched on and off in the embryology of higher organisms? Third, the molecular analysis of bacteriophage has revealed con-

siderable detail about the intracellular synthesis of viral DNA and proteins. The control of the assemblage of new virus particles from these components involves both genetic and nongenetic steps. The experiments of Edgar and Epstein reveal how these events lead to new virus formation and, by inference, they point out that this system could serve as a molecular model for cell differentiation.

H. V. Wilson

On Some Phenomena of Coalescence and Regeneration in Sponges

The disaggregation of sponges into their individual cells does not destroy their capacity for recognition and adhesion. New sponges can be formed by the proper orientation of these tissues. The phenomenon of cell movement leading to tissue organization is called *morphogenetic movement.* Once Wilson achieved regeneration of sponges from dissociated cells of the sponge, *microciana,* he repeated his techniques on other sponges (e.g., *Stylotella*) and attempted interspecific mixtures of dissociated cells to see if the cell recognition was specific or not. He found that the cells were species specific. Wilson's viewpoint is that the organism is not merely a "whole" unit but that its parts may be put to experimental analysis without fear that the process destroys the integrity of the organism. The organism-as-a-whole viewpoint, if taken literally, would prevent any meaningful analysis of cells or their parts because the detachment would lead to a loss of the "integrity" of the organism. Modern biologists usually reject the organism-as-a-whole outlook, but they are aware that higher organisms do have a complexity that may be destroyed by analytic laboratory treatment. For this reason the experimentalist tries to design his work carefully so that the phenomenon he finds is not a coincidental artefact of his analytical technique.

I

In a recent communication I described some degenerative and regenerative phenomena in sponges and pointed out that a knowledge of these powers made it possible for us to grow sponges in a new way. The gist of the matter is that silicious sponges when kept in confinement under proper conditions degenerate in such a manner that while the bulk of the sponge dies, the cells in certain

Reprinted from *The Journal of Experimental Zoölogy V*:245–258, 1907.

regions become aggregated to form lumps of undifferentiated tissue. Such lumps or plasmodial masses, which may be exceeding abundant, are often of a rounded shape resembling gemmules, more especially the simpler gemmules of marine sponges (Chalina, e.g.), and were shown to possess in at least one form (Stylotella) full regenerative power. When isolated they grow and differentiate producing perfect sponges. I described moreover a simple method by which plasmodial masses of the same appearance could be directly produced (in Microciona). The sponge was kept in aquarium until the generative process had begun. It was then teased with needles so as to liberate cells and cell agglomerates. These were brought together with the result that they fused and formed masses similar in appearance to those produced in this species when the sponge remains quietly in aquarium. At the time I was forced to leave it an open question whether the masses of teased tissue were able to regenerate the sponge body.

During the past summer's work at the Beaufort Laboratory[1] I again took up this question and am now in a position to state that the dissociated cells of silicious sponges after removal from the body will combine to form syncytial masses that have power to differentiate into new sponges.[2] In Microciona, the form especially worked on, nothing is easier than to obtain by this method hundreds of young sponges with well-developed canal system and flagellated chambers. How hardy sponges produced in this artificial way are and how perfectly they will differentiate the characteristic skeleton, are questions that must be left for more prolonged experimentation.

Taking up the matter where it had been left at the end of the preceding summer, I soon found that it was not necessary to allow the sponge to pass into a degenerative state, but that the fresh and normal sponge could be used from which to obtain the teased-out cells. Again in order to get the cells in quantity and yet as free as possible from bits of the parent skeleton, I devised a substitute for the teasing method. The method adopted is rough but effective.

Let me briefly describe the facts for Microciona. This species (*M. prolifera Verr.*) in the younger state is incrusting. As it grows older it throws up lobes and this may go so far that the habitus becomes bushy. The skeletal framework consists of strong horny fibers with embedded spicules. Lobes of the sponge are cut into small pieces with scissors and then strained through fine bolting

cloth such as is used for tow nets. A square piece of cloth is folded like a bag around the bits of sponge and is immersed in a saucer of filtered sea-water. While the bag is kept closed with the fingers of one hand it is squeezed between the arms of a small pair of forceps. The pressure and the elastic recoil of the skeleton break up the living tissue of the sponge into its constituent cells, and these pass out through the pores of the bolting cloth into the surrounding water. The cells, which pass out in such quantity as to present the appearance of red clouds, quickly settle down over the bottom of the saucer like a fine sediment. Enough tissue is squeezed out to cover the bottom well. The cells display amoeboid activities and attach to the substratum. Moreover they begin at once to fuse with one another. After allowing time for the cells to settle and attach, the water is poured off and fresh sea-water added. The tissue is freed by currents of the pipette from the bottom and is collected in the center of the saucer. Fusion between the individual cells has by this time gone on to such an extent that the tissue now exists in the shape of minute balls or cell conglomerates of a more or less rounded shape looking to the eye much like small invertebrate eggs. Microscopic examination shows that between these little masses free cells also exist, but the masses are constantly incorporating such cells. The tissue in this shape is easily handled. It may be sucked up to fill a pipette and then strewn over cover glasses, slides, bolting cloth, watch glasses, etc. The cell conglomerates which are true syncytial masses throw out pseudopodia all over the surface and neighboring conglomerates fuse together to form larger masses, some rounded, some irregular. The details of later behavior vary, being largely dependent on the amount of tissue which is deposited in a spot, and on the strength of attachment between the mass of tissue and the sub-stratum.

Decidedly the best results are obtained when the tissue has been strewn rather sparsely on slides and covers. The syncytial masses, at first compact and more or less rounded, flatten out becoming incrusting. They continue to fuse with one another and thus the whole cover glass may come to be occupied by a single incrustation, or there may be in the end several such. If the cover glass is examined at intervals, it will be found that differentiation is gradually taking place. The dense homogeneous syncytial mass first develops at the surface a thin membrane with underlying connective tissue (collenchyma). Flagellated chambers make their

appearance in great abundance. Canals appear as isolated spaces which come to connect with one another. Short oscular tubes with terminal oscula develop as vertical projections from the flat incrustation. If the incrustation be of any size it produces several such tubes. The currents from the oscula are easily observed, and if the cover glass be mounted in an inverted position on a slide the movements of the flagella of the collar cells may be watched with a high power (Zeiss 2 mm.). This degree of differentiation is attained in the course of six or seven days when the preparations are kept in laboratory aquaria (dishes in which the water is changed answer about as well as running aquaria). Differentiation goes on more rapidly when the preparation is hung in the open harbor in a live-box (a slide preparation inclosed in a coarse wire cage is convenient). Sponges reared in this way have been kept for a couple of weeks. The currents of water passing through them are certainly active and the sponges appear to be healthy. In such a sponge spicules are present, but some of these have unquestionably been carried over from the parent body along with the squeezed out cells.

The old question of individuality may receive a word here. Microciona is one of that large class of monaxonid sponges which lack definite shape and in which the number of oscula is correlated simply with the size of the mass. While we may look on such a mass from the phylogenetic standpoint as a corm, we speak of it as an individual. Yet it is an individual of which with the stroke of a knife we can make two. Or conversely it is an individual which may be made to fuse with another, the two forming one. To such a mass the ordinary idea of the individual is not applicable. It is only a mass large or small having the characteristic organs and tissues of the species but in which the shape of the whole and the number of the organs are indefinite. As with the adult so with the lumps of regenerative tissue. They have no definiteness of shape or size, and their structure is only definite in so far as the histological character of the syncytial mass is fixed for the species. A tiny lump may metamorphose into a sponge, or may first fuse with many such lumps, the aggregate also producing but a single sponge although a larger one. In a word we are not dealing with embryonic bodies of complicated organization but with a reproductive or regenerative tissue which we may start on its upward path of differentiation in almost any desired quantity. A striking illustration of this nature of the material

is afforded by the following experiment. The tissue in the shape of tiny lumps was poured out in such wise that it formed continuous sheets about one millimeter thick. Such sheets were then cut into pieces, each about one cubic millimeter. These were hung in bolting cloth bags in an outside live-box. Some of the pieces in spite of such rough handling metamorphosed into functional sponges.

Even where the embryonic bodies of sponges have a fixed structure and size, as in the case of the ciliated larva, the potential nature as displayed in later development, is not fixed in the matter of individuality. Such a body may form a single individual or may fuse with some of its fellows to form a larger individual differing from the one-larva sponge only in size. It is then, in spite of its definiteness of shape and size, essentially like a lump of regenerative tissue in that whether it develops into a whole sponge or a part of a sponge depends not on its own structure but on whether it is given a good opportunity of fusing with a similar mass. A parallel case to the coalescence of larvae is afforded by the gemmules of fresh-water sponges. Mr. M. E. Henriksen in a manuscript account submitted to me a year ago, describes the fusion of gemmules to form a single sponge.

In the preceding description I have passed over the question as to the precise nature of the cells which combine to form the masses of regenerative tissue. On this point as on the histological details in general I hope to have more to say later. Nevertheless the phenomena are so simple that observation of the living tissue reveals much, probably indeed all that is of fundamental importance. If a fairly dense drop of the squeezed-out tissue be mounted at once and examined with a high power (Zeiss 2 mm., comp., oc. 6), the preparation is seen to consist of fluid (sea-water) with a few spicules and myriads of separate cells. The cells fall into three classes.

1 The most conspicuous and abundant are spheroidal, reddish, densely granular, and about 8 μ in diameter. These cells, which can be nothing but the unspecialized, amoeboid cells of the mesenchyme (amoebocytes or archaeocytes), put out hyaline pseudopodia that are sometimes elongated, more often rounded and blunt.

2 There is also a great abundance of partially transformed collar cells, each consisting of an elongated body with slender flagellum. The cell is without a collar, the latter doubtless having

been retracted. In the freshly prepared tissue the flagella are vibratile, the cells moving about. Soon however the flagellum ceases to vibrate.

3 The third class is not homogeneous. In it I include more or less spheroidal cells ranging from the size of the granular cells down to much smaller ones. Many of these are completely hyaline, while others consist of hyaline protoplasm containing one or a few granules.

Fusion of the granular cells begins immediately and in a few minutes time most of them have united to form small conglomerate masses which at the surface display both blunt and elongated pseudopodia. These masses soon begin to incorporate the neighboring collar and hyaline cells. One sees collar cells sticking fast by the end of the long flagellum to the conglomerate mass. Other collar cells are attached to the mass by short flagella. Still again only the body of the collar cells projects from the mass while there is no sign of the flagellum. Similarly spheroidal hyaline cells of many sizes are found in various stages of fusion with the granular conglomerate. In such a preparation the space under the cover glass is soon occupied by innumerable masses or balls of the kind just described, between which continue to lie abundant free cells, some collar cells, others hyaline. Practically all the granular cells go to make up the balls. The play of pseudopodia at the periphery of such balls, which results in the incorporation of free cells and in the fusion of balls to form larger masses, is easily watched. Along with such a cover-glass preparation it is convenient to have some of the squeezed-out tissue in a watch glass of sea-water. In the watch-glass preparation it is instructive to watch with a two-thirds or one-half objective the fusion of the cell conglomerates to form masses like those strewn on covers, slides, etc.

These observations on the early steps in the formation of the masses of regenerative tissue make it plain that such masses are composed chiefly of the spheroidal, granular cells (amoebocytes or archaeocytes), but that nevertheless other cells, collar cells and more or less hyaline cells also enter into their composition. I may recall the fact that in the formation of regenerative masses in a degenerating sponge,[3] the evidence from sections, which is the only evidence available in the case, points to the conclusion that the collar cells help to form the syncytial tissue of the masses. The question of interest lying at the heart of this matter may be so

formulated: can particles of the Microciona protoplasm differentiate into functional collar cells and, when the occasion arises, change back into unspecialized masses capable of combining with other masses of unspecialized protoplasm to form a regenerative body? The facts to which I have just alluded support this idea, and indicate that the immediate problem is one worth pursuing farther as a good case of temporary differentiation of protoplasm in the metazoa analogous to the temporary specialization of the cell individual which occurs in such colonial protozoa as Protospongia.[4]

As far as the amoebocytes are concerned it is certain that they have great regenerative power. Weltner in a recent paper[5] has emphasized the importance of these unspecialized cells in the processes of growth and regeneration. His conclusions which refer directly to fresh-water sponges, are that in a growing sponge, in a sponge regenerating new organs after its winter period of simplification, and in the regeneration of a sponge from a cutting, the amoebocytes are the all-powerful elements in that they give rise to all the new tissues formed. He further alludes to the fact that such reproductive bodies as the gemmules of fresh-water sponges and the buds of Tethya (according to Maas) are only groups of amoebocytes; further that the gemmules of Tedania and Esperella described by Wilson as developing into ciliated larvae, and the similar bodies found by Ijima in hexactinellids, are such groups. I may add that the presence of such groups of unspecialized cells in the hexactinellids has recently been confirmed by the master in sponge-morphology, F. E. Schulze, who recognizes the probability of their reproductive nature and gives them a new name, that of *sorites*.[6] It is clear then that in many sponges reproductive bodies are formed by the association of unspecialized amoeboid cells. But there is nothing in this fact which precludes the possibility that the groups of amoebocytes are in part recruited from transformed collar cells and other tissue cells, such as pinacocytes (flat cells of canal walls), that have undergone regressive differentiation into an unspecialized amoeboid condition.

Cells analogous to the amoebocytes of sponges are found elsewhere in the metazoa, e.g., in the ascidians.[7] It would be interesting to know what capacity, if any, for development they have, when freed from the parent (bud) and collected together in seawater.

II

I shall here briefly record some experiments which gave only negative results but which under circumstances admitting of a wider choice of species, ought to yield returns of value. These experiments were based on the assumption that if the dissociated cells of a species will recombine to form a regenerative mass and eventually a new sponge, the dissociated cells of two different species may be made to combine and thus form a composite mass bearing potentially the two sets of species-characteristics. It is clear that such an organism would be analogous to one produced by an association of the blastomeres of the two species. Pending the successful carrying out of this experiment, it would be idle to discuss further the nature of the hypothetical dual organism.

In my own experiments three sponges were used: Microciona, Lissodendoryx and Stylotella. The three are all monactinellids, but Microciona is the only one in which the skeleton includes any considerable amount of horny substance. Dissociated cells of Microciona and Lissodendoryx were mixed, and again dissociated cells of Microciona were mixed with those of Stylotella. In each case the experiment was performed at two different times, and a considerable number of admixtures, in watch glasses and on cover glasses, was made. The preparations were examined at short intervals with the microscope. The cells of these three species are colored very differently, and are therefore easily distinguished, at least as soon as fusion sets in and little masses of cells begin to be formed. In all the experiments the cells and cell-masses of a species combined, and not the cells of different species. Thus in the admixture of Microciona and Lissodendoryx, Microciona regenerative masses and Lissodendoryx regenerative masses were produced. Similarly when Microciona and Stylotella cells were mixed, the resultant masses were pure, some Microciona, some Stylotella. The Microciona masses in these experiments were hardy. They continued to develop and in some preparations metamorphosed. The cell masses of the other two species, while they reached a considerable size were not hardy, most dying soon although some began the process of metamorphosis.

These three species are so unlike that there was little ground in the beginning for the expectation that coalescence would take place. Possibly, as in the cases where fusion of egg and sperm of different species is induced through some alteration in the physi-

ological state of the protoplasm, so the regenerative cells and cell masses of different species may be made to combine under abnormal conditions. The more promising task is, however, to find allied species and sub-species, the regenerative tissue of which will combine under natural conditions. Such forms, I take it, should be sought among the horny sponges and monactinellids with abundant horny matter.

III

The tendency to fuse so vigorously displayed by the cells and cell masses of regenerative tissue led me to examine into the power that larvae have to fuse with one another and the capacity for development in the resultant mass. Delage and others have remarked on the not infrequent occurrence of fusion between sponge larvae. Delage[8] says that he has often observed two or several larvae unite to form a single sponge "which has from the start several cloacas."

I find that this power to fuse displayed by the larvae is one that is easy to control. Fusion between larvae will readily take place if they are brought in contact at the critical time when the ciliated epithelium is being replaced by the permanent flat epithelium. At this time they will fuse in twos or threes or in larger number up to and over one hundred (Figs. 1–4). The smaller composite masses composed of as many as five or six larvae metamorphose into perfect sponges. The larger masses composed of many larvae did not metamorphose in my experiments but experience with the regenerative tissue suggests that such masses would metamorphose if certain mechanical difficulties due to the great size of the mass were removed. Possibly this might be accomplished by cutting a flattened sheet composed of some hundred larvae (such as I have produced) into pieces and inducing the pieces to metamorphose separately.

I may now describe some of the details in this process of larva-fusion. In a species of Lissodendoryx used the larva is of the following character. It has the usual ovoidal shape with a posterior protuberant non-ciliated pole. The anterior pole is somewhat truncated and is sparsely ciliated. The rest of the body bears the usual thick covering of cilia. As seen with reflected light the bulk of the body is dead white, the posterior pole deep blue, and the anterior pole bluish. This coloration is not absolutely fixed for the species, but the larvae used in my coalescence experiments

were all of this character. Within twenty-four hours after liberation the ciliated larvae are creeping (remaining in contact with the bottom as they swim) over the bottom of the dish. Some are now put in deep round watch glasses and with pipette and needle coaxed together into a clump. Fusion soon begins and on the next day plenty of composite larvae are present. The larvae fuse endwise, for the most part in pairs. The compound larva so produced owing to its weight has a very feeble locomotory power. Using pairs that are nearly motionless, larvae may be brought together (coaxed with needle) and arranged in a desired position on a cover glass for instance. In successful cases fusion results before the separate masses move apart. In this way, selecting an instance, I have added to one arm of a quadruple mass a pair of larvae, and to the opposite arm two pairs (Fig. 4).

For the purpose of bringing about the fusion of many larvae the following simple method is convenient. Suppose that we have the larvae in a paraffine-coated dish, and they are in a late "creeping" stage. Small excavations, 2–3 mm. deep and 4–5 mm. wide, are now made in the paraffine, and with the pipette the larvae are driven into the holes. They lie here in numbers up to and over one hundred, crowded together and heaped upon one another. Fusion begins soon and the larvae are gradually converted into a flattened cake. The larger cakes thus made measured four by three millimeters. The body of such a cake is a continuous flattened mass in which there is no indication of the component larvae, but the rounded ends of the larvae that have last fused with the general mass remain for a time distinguishable. Owing to their blue coloration the ends of the larvae may be recognized in these and the other compound masses even after the outline of the larva has been completely lost.

As already stated the smaller compound masses metamorphose without difficulty. The coalesced larvae may be made to attach to cover glasses, slides, etc. Larger masses composed of about twenty larvae underwent a partial metamorphosis. Such masses were laid upon bolting cloth to which they readily attached. The largest masses were hung in small bolting-cloth bags in a live-box. Whether owing to bad handling or more probably to some inherent difficulty, they did not metamorphose but soon died.

The ease with which larvae of the same species may be made to fuse together suggests that larvae of different species might

likewise be induced to coalesce. Some experiments along this line could not fail to be of interest.

IV

In the tendency to fuse with the production of a plasmodium, the dissociated cells of sponges resemble the amoebocytes (amoebulae) of the mycetozoa and Protomyxa. The regenerative power of the plasmodium has an interest both theoretical and economic in itself. But it is the tendency to fuse displayed by the cells that have been forcibly broken apart, which constitutes the fact of most general physiological importance. Discarding for the moment the word "cell" and speaking of the protoplasm of a species as a specific substance, the phenomena may be restated to advantage in the following way.

A mass of sponge protoplasm in the unspecialized state typically exhibits pseudopodial activities at the surface. In lieu of more precise knowledge it is useful to regard the pseudopodia as structures which explore and learn about the environment. On coming in contact two masses of the same specific protoplasm tend to fuse. This tendency is probably useful (i.e., adaptive) in that the additional safety (from enemies and "accidents") accruing from increase in size of the mass more than compensates for the reduction in number of the individual masses that start to grow (rearing of sponges shows that masses of good size frequently withstand conditions that effectually wipe out the very small masses). Unlike specific substances (protoplasm of quite different species) do not tend to fuse.

To the many biologists who have found ideas and observations of deep interest in the papers on protoplasmic activities by Professor and Mrs. E. A. Andrews (G. F. Andrews), the statement just made will have a familiar sound. Mrs. Andrews in her essay on The Living Substance as Such and as Organism[9] and her paper on The Spinning Activities of Protoplasm[10] makes, it would appear from subsequent confirmations, a definite advance in our knowledge of the intimate structure of protoplasm. But it is her generalizations, based on singularly acute observations, with respect to the *behavior* of protoplasm, that have especially influenced my own work. The particular generalizations referred to may be so formulated:

1 Protoplasm tends to produce a viscous, pellicular layer with formation of pseudopodial outgrowths over the surface, whether

external or internal to the mass, which establishes contact with the environmental medium.

2 Pseudopodia from adjacent masses of the same specific substance tend to fuse. Thus actual connections which can be made and remade, and along which transference of substance takes place, are established between the masses.

That these phenomena are observable in widely separated groups of metazoa has been also shown by Professor Andrews in a series of brief studies marked with his well-known skill and accuracy of observation and statement. I fully agree with him as to the great importance of the facts.

The general point of view entertained by Mrs. Andrews in her much discussed essay is perhaps not everywhere clear to me. It is manifest, however, that she consistently subordinates the idea of the individual, whether entire organism or cell, to that of the specific substance of which it is but a more or less detached piece. As far as the cell is concerned this point of view seems to be essentially that of Sachs and Whitman. Mrs. Andrews extends it to the whole organism, and I may say that this way of looking at an animal or plant (or piece of the same) is in my opinion a habit of mind that will justify itself and indeed is doing so today, in that it leads to discoveries concerning the nature of protoplasms as revealed by what they can do.

1907

NOTES

[1] I am indebted to the director of the station, Mr. H. D. Aller, for his kindly aid in supplying all facilities needed in the course of my investigation.

[2] These findings are being published with the permission of Hon. Geo. M. Bowers, U.S. Commissioner of Fisheries.

[3] A new method by which sponges may be artifically reared, *Science,* n.s., vol. XXV, no. 649, 1907.

[4] Metschnikoff, *Embryologische Studien an Medusen,* p. 147, 1886.

[5] "Spongilliden-studien V. Zur Biologie von Ephydatia fluviatilis und die Bedeutung der Amoebocyten für die Spongilliden." *Archiv für Naturgeschichte,* 73 Jahrg., 1 Bd., 2 Heft, 1907.

[6] *Wissensch. Ergebn. d. Deutsch. Tiefsee-Exp.* 1898–99. "Hexactinellida," pp. 213–15. Jena, 1904.

[7] Comp. Hjort's and LeFevre's papers on budding in ascidians.

[8] "Embryogénie des Éponges." *Arch. de Zool. Exp. et Gén.,* p. 400, 1892.

[9] Suppl. to *Journ. Morphology,* vol. XII, no. 2, 1897.

[10] *Journ. Morphology,* 1897.

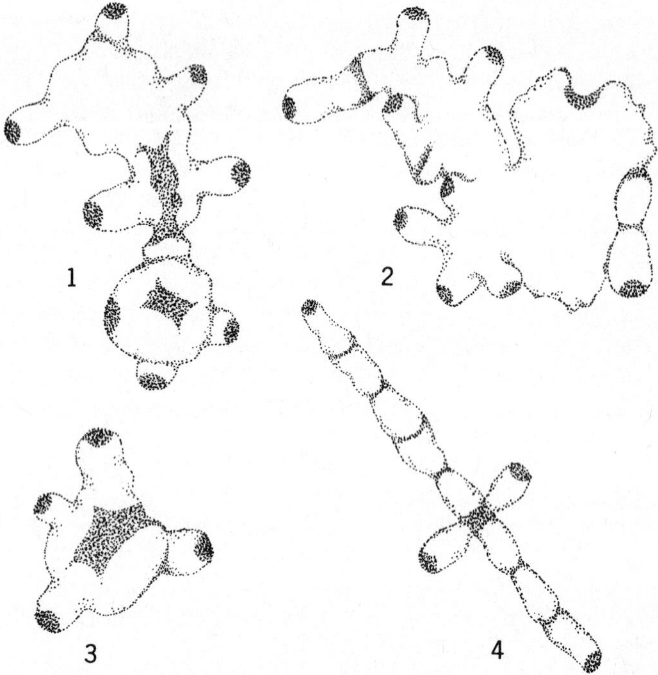

FIGS. 1, 2, 3, 4. Composite masses produced by the fusion of larvae, The stippled ends and areas are in nature blue, and represent the ends of the component larvae. The body of the mass is white. Fig. 1 shows a mass composed of four larvae which has just united with a mass composed of five or six larvae. In Fig. 2 more than ten, probably about twenty, larvae have combined. In Fig. 3 about six larvae have combined. In Fig. 4 the original quadruple mass composed of four radiately arranged larvae, has been extended in one direction by the addition of a pair of larvae, and in the opposite direction by the addition of two pairs of larvae. Figs. 1 and 3 ×44; Figs. 2 and 4 ×22.

Hans Spemann

Embryonic Development and Induction

The importance of the cell nucleus was recognized in the late nineteenth century. The nucleus was acknowledged as the bearer of hereditary material because the sperm were virtually void of cytoplasm and they contributed as much to the inheritance of the progeny as did the eggs. But after fertilization, according to Weismann, some cells are set aside as germ plasm and other cells differentiate to form somatic tissues. Were the nuclei of progressively older cells identical to the nuclei of the fertilized egg? Was somatic cell differentiation accompanied by changes in the genotype of the nucleus? Spemann attempted to answer this by using a fine hair to ligature a fertilized egg of the salamander, Triton. He allowed the nucleated half to multiply and then let a nucleus slip into the anucleate half. The embryo formed was normal.

In the early 1950's a new method was devised to extend Spemann's work. R. Briggs and R. King used fine glass needles to remove nuclei from the fertilized eggs of the leopard frog, *Rana pipiens.* They replaced the nuclei with cells removed from older embryos. Even in later stages of development (gastrulae) some of the nuclei of differentiated tissues could generate a normal embryo. In many other instances, however, the nuclei of differentiated tissues produced abnormal embryos. The basis for the abnormalities is unknown. The problem of nuclear differentiation has not been resolved, but the nucleus retains its potential for normal development for longer periods of time than it was once believed possible.

Constricting experiments on Triton's eggs. O. Hertwig (1893), stimulated by Roux's isolation experiment, made an experiment which, although unsuccessful, became of great significance because it employed a new method for the first time and introduced an eminently suitable experimental object. He constricted eggs (in the two-cell stage) of the common newt (*Triton taeniatus*) with a

Reprinted from *Embryonic Development and Induction* (New Haven: Yale University Press, 1938), pp. 22–39, by permission of the publisher and the author.

hair loop along the first cleavage plane in order, if possible, to separate completely the first two blastomeres. As I said before, a satisfactory result was not obtained, probably because of an insufficient number of experiments. A few years later, however, Endres (1895) and Herlitzka (1897) succeeded by this method in producing twins from the separated blastomeres; the twins were well-developed embryos of half size but of normal proportions and corresponded completely with those produced by Driesch from sea urchin's eggs. Later on this experiment was taken up by me (1901–1903) and carried farther on a larger scale (Fig. 1). Besides a confirmation of the older results it was shown that, with a less effective constriction by which the blastomeres are not completely separated, partial duplications could be obtained. These either involve the fore end only (duplicitas anterior) or the whole germ, but the latter in the shape of a cross-wise concrescence for which I later proposed the term *"duplicitas cruciata."*

Isolation experiments on frog's eggs. Finally, it was shown by a variety of experiments that the same holds true for Roux's own object, the frog's egg.

If the two first blastomeres of the frog's egg are completely isolated, or if, having pricked one of them, we make the other completely independent, we obtain twins or a whole embryo of half size, as in the case of *Triton's* eggs. In the latter experiment the cell which had been killed by pricking was either sucked off (McClendon, 1910) or altogether separated, by pressing, from the surviving one which now continued to develop as a whole. The separation of the two living halves of the germ was quite recently done in my laboratory by G. A. Schmidt (1933) by constriction with a fine fiber of artificial silk. Earlier experiments by O. Schultze (1894) and T. H. Morgan (1895), in which formation of a whole was attained by a rearrangement of the egg substance, will be discussed later.

Delayed nucleation. In the above experiments, a blastomere, the nucleus of which, according to Weismann, should contain the determinants for half an embryo only, has furnished a whole one. It is still more striking, however, when the same potency may be demonstrated for still smaller fragments of the fertilized egg nucleus, by delaying the nucleation of one half of the egg. This experiment also has been made on the eggs of sea urchins and amphibians with equal success.

J. Loeb (1894) placed fertilized eggs of sea urchins in diluted

sea water for some time. In consequence of absorption of water from the hypotonic medium, the egg membrane bursts and a portion of an egg plasm protrudes as in a hernia. If this protruding plasm does not contain a nucleus, it is at first excluded from development. Not until a few divisions of the nucleus in the half of the egg containing it have taken place, does a secondary nucleus wander across the plasma bridge into the other half and initiate its development. According to Weismann this nucleus should contain a fraction of the germ plasm, and, consequently, allow a part only of the embryo to arise. Instead of this, a whole formation is developed or half of a double monster which is connected with the other part remaining inside the membrane.

Exactly the same results may be obtained with the eggs of *Triton* (Spemann, 1914, 1928; Schütz, 1924; Frankhauser, 1925, 1930) which have been constricted by a hair loop shortly after fertilization. Here also at first only the half containing the nucleus begins to cleave (Fig. 2) while the other half remains undeveloped (Fig. 3) until, according to the degree of tightness of the ligature, an earlier or a later descendant of the cleavage nucleus wanders across the broad (Fig. 4) or narrow (Fig. 5) bridge of plasm and provides the other half of the egg with a nucleus. This half now develops as if it possessed a whole cleavage nucleus; hence, if the constriction lay in a median plane, it develops into an embryo (Fig. 6, *a* and *b*) which either can be independent or united with the other in a double formation.

The result is particularly impressive when the tying takes place not in an approximately median (Fig. 7, *a*), but in an approximately frontal plane (Figure 7, *b*). The constriction, partial or complete, of a two-cell stage produced different results according to whether the first cleavage plane was median or frontal or approximately in one of these directions (Spemann, 1901–1903). While in the first case, according to the degree of tightness of the ligature, double formations or twins resulted, in the second a dorsal germ half with axial organs was separated from a ventral half. This separation is incomplete with loose tying; in this case the ventral half hangs from the dorsal half like a yolk sac. When the tying is tight or when the two halves are completely separated, the dorsal half develops into a small embryo of normal proportions (Fig. 8, *a*). The ventral half, however, produces a ventral part only (Fig. 8, *b*), composed of the three germ layers, but without any axial organs. Now, such different products can also result

after constriction of the nonsegmented egg. In this case, the half of the germ which only contains a fraction of the cleavage nucleus can produce the whole embryo while the other half, in spite of the almost complete nucleus, forms a ventral part only. The course of development that a part of the germ will take obviously depends on the egg plasm, not on the nucleus (Spemann, 1914, 1928).

Ventral parts of the same type as well as other symmetrical and asymmetrical defective formations occurred also in experiments of Frankhauser (1930*b*) in eggs which had been constricted completely ten to fifty minutes after fertilization. Frankhauser is surely justified in concluding that at this early stage the typical distribution of the egg substances is established. These remarkable results were brought by him into agreement with those of Brachet (1906), who observed circumscribed defects in frog embryos after having pricked the unsegmented eggs.

Abnormal distribution of the nuclei during segmentation. By still another method the normal distribution of nuclei may be interfered with, which according to Weismann's theory would necessarily have led to an abnormal development of the regions affected.

When an egg of a sea urchin or an amphibian is strongly compressed between two parallel glass plates, the first cleavage planes, instead of at first forming right angles and later variable angles with each other, come now to stand all perpendicularly to the compressing surfaces and, instead of a globular mass, we have at first one single layer of cells. The cell nuclei also thus come to lie in one and the same plane; hence they will obviously be placed in abnormal parts of the egg plasm. If, now, it is assumed that the nuclei differ from each other because of differential division, and if they later on determine the further development and the ultimate fate of their respective cells, as Weismann's theory postulates, then the parts of the developed germ would have to be displaced in an abnormal manner. Instead of this, perfectly normal embryos resulted after the compression was relieved. This has been proved for the eggs of sea urchins by H. Driesch (1892) and for those of frogs by G. Born (1893) and O. Hertwig (1893).

Last but not least, Boveri (1910) has raised grave objections to Weismann's theory from quite another angle. He showed that even if a differential division did exist, the mitotic apparatus of the cell would not possess the power to ensure a correct distribu-

tion of the two different halves of chromosomes between the two new descendant cells.

Criticism of Weismann's theory. Weismann's theory is not compatible with the above facts or becomes so only if additional hypotheses are introduced which, however, deprive it of any explanatory value. Besides the activated idioplasm which is assigned to each kind of cell by the differential division, there is supposed to be some still undivided idioplasm, "reserve idioplasm," which is distributed for special emergencies in which, in the case of the above experiment, it would be supposed to direct the formation of a whole. But besides the fact that such an hypothesis, invented *ad hoc* as it is (Driesch called it a "photograph of the problem") possesses but little explanatory value, a new difficulty is presented by this reserve idioplasm. If it is to bring about that which is required, it cannot possibly be distributed from merely internal causes, but must be activated to meet the requirements of the special situation. In other words, just those capabilities must be ascribed to the idioplasm which the theory declined to ascribe to the original undivided germ plasm, although the reserve idioplasm is supposed to be identical with the latter. This part of Weismann's germ-plasm theory has been generally abandoned.

But beyond this negative result, the facts described have stimulated research, raising problems of a quite new order. But before this matter may be treated in detail, we must first discuss the apparently contradictory results of the pricking and the constriction experiments.

Discussion of the different results of pricking and complete separation. It has already been mentioned that the difference in result is not due to the different material of the experiments. Not only in the egg of the sea urchin and *Triton,* but also in that of the frog, a whole embryo originates from a completely isolated blastomere. That same blastomere, however, forms—at least at first—a half embryo when the other blastomere, although destroyed, remains in connection with it. This is the case not only in the frog's egg, but also in the egg of the axolotl, as we know from the work of D. Barfurth (1893). Hence the difference in the method must be responsible for the different result. The tendency toward half formation as well as the potency to form a whole is present in both kinds of eggs in the same manner within the half blastomere. But when the isolation is absolute, a change toward

the "whole," a regulation, takes place. When the isolation is incomplete this change does not occur. The cause of this may be deduced from an experiment of O. Schultze and T. H. Morgan.

A frog's egg, in the two-cell stage, was compelled to maintain an inverted position, after having been previously compressed vertically to the egg's axis (O. Schultze, 1894) or after pricking one blastomere (T. H. Morgan, 1895). Each of the two blastomeres in the first case, or the surviving one in the second case, developed into a whole embryo. In Schultze's experiments the embryos were connected in different ways, i.e., double formations of different kinds resulted. Schultze's experiment was repeated by G. Wetzel (1895) and, after him, a number of times, most thoroughly by W. Schleip and A. Penners (1925, 1926). These experiments suggested that the stratified egg substances, being of different weight, began after inversion to flow in different ways, due to minor accidental influences; and because of their rearrangement double and multiple monsters resulted.

The actual facts may perhaps be considered to be as follows. The fertilized egg possesses a bilaterally symmetrical structure, which, according to Roux, is established by the entrance of the sperm and the ensuing flowing of the egg plasm. This structure is transferred to the first blastomeres, each of which has, therefore, the structure of one half. The killing of the one blastomere does not destroy the "half" structure of the other. Just as a bilaterally symmetrical embryo arises because of the bilaterally symmetrical structure of the whole egg, so a half embryo arises out of a half egg (cf. p. 19; W. Vogt, 1927, 1928 *a* and *b*). But upon the complete removal of one of the blastomeres, an internal rearrangement of structure takes place in the remaining one, which can also be brought about by an artificially created flow. There follows a regulation of the "half" to the "whole," so that a bilateral twin embryo results.

Driesch's harmonious-equipotential system. Driesch has, in addition to criticizing Weismann's theory, also made positive deductions from the results of his experiments. He conceives the segmenting egg of the sea urchin—and the same would also apply to the amphibian egg—as a system of parts which are all able to perform alike, which all possess the same *potency* or are "equipotential," and thus form an "equipotential system." This system of equivalent parts now subdivides itself harmoniously (it is "harmonious-equipotential") according to an inherent proportion

into smaller systems in which—as Driesch concluded from later experiments—the same process repeats itself. It divides into harmonious-equipotential subsystems, until the proper rôle in development has been assigned to every part of the embryo. The frame, or the system of coördinates, so to speak, of this process is a very general bilaterally symmetrical microstructure of the egg which, as in a magnet, repeats itself in its smallest parts. With regard to the structure of the embryo the fate of its parts is a "function of their position in the whole."

This kind of development according to Driesch holds true not only in the case of the experiment, that is, after disturbance, but also in the normal, undisturbed development. The logical principle of economy underlies this very important statement. If we are able to prove the presence—in the germ—of an ability which suffices for the explanation of the normal development, we are not justified in adopting a further principle unless forced to do so by special facts.

Roux's conception of primary and secondary development: postgeneration. Here we must speak of one other result of Roux's pricking experiment, the far-reaching significance of which was fully recognized by Roux himself. He found that the half embryo which had developed from the surviving half of the egg completes itself later on, making a whole embryo, either out of its own substance by internal transformation or by utilization of the other half, if this was only damaged instead of killed outright by the puncture. Roux compared this process with regeneration and called it, in distinction from it, "postgeneration." There has been much discussion about the details of this phenomenon; indeed, even its actual occurrence has been called in question. The doubts were caused by the great difficulties of accurate observation which were necessarily connected with Roux's method, but I believe, chiefly by reason of experiments of my own by another method that, in the main, Roux saw and interpreted correctly.

The same phenomena can be easily produced, by a somewhat different method, also in the egg of *Triton,* and here they can be followed with much greater exactness. When a median separation of the germ is performed not in the two-cell stage, but later at the beginning of gastrulation, a regulation toward the whole also takes place, but it is, so to speak, overtaken by the advancing development. Thus, the twin embryos develop a weaker formation of the inner side and this may go so far as to produce the same phenomena observed in the postgeneration of Roux's sur-

viving frog blastomeres out of their own material. The cells of the surviving half germ, in the pricking experiment, are probably kept in their normal position at first by the adhering dead mass and only separate from it because of the more extensive displacements which set in at the time of gastrulation. Then they close and thus receive the stimulus toward regulation. Hence also in the pricking experiment the isolation proper of the surviving half of the germ would, according to this interpretation, occur at the beginning of gastrulation only, and the regulative power of the germ would be the same in the case of the frog as in that of the newt (Spemann and Falkenberg, 1919).

Discussion of some conceptions and terms. If the facts which have been frequently verified can, without being overstrained, be joined together in one harmonious entire picture, the conceptions in which these facts have found their expression cannot contradict each other. This is indeed the case. The conception of the harmonious-equipotential system is incompatible with Weismann's theory in its pure formulation only, but not with the conception of self-differentiation, nor even necessarily with the mosaic theory.

Self-differentiation, according to Roux's definition, always refers to a sharply circumscribed region in a definite stage of development, which, in this case, is one of the halves of the egg in the two-cell stage. The independence of this region of external influences by no means implies that its separate parts also must develop independently of each other. R. G. Harrison (1918) once expressed this poignantly by calling the rudiment of the limb, a "self-differentiating harmonious equipotential system."

When a part of an embryo carries within itself the causes for its further development in a definite direction, it may be said that it is predestined for its future fate or "determined." In any case one can define with F. R. Lillie (1929) the conception of "determination" in such a way that one makes the power of self-differentiation its criterion. Then the lateral half blastomere of the frog's egg would be *determined* to form a lateral half embryo. This example shows that "determination," thus interpreted, need not be irrevocable, but that it can be canceled and take place in another direction.

The mosaic theory adopts this determination for the minutest parts of the germ. If here also determination were not irrevocable but "labile," even the mosaic theory would not necessarily contradict the conception of the harmonious-equipotential system,

because this term only implies that the single parts of the embryo still possess the same potency. It does not preclude the possibility that they are already determined in a certain direction. Yet it is true that in this case, normal development would take a different course from the disturbed one. The experiment might reveal "capabilities" of the embryo, but would not tell us anything about the part which they play in normal development. This is indeed the position Roux took against Driesch.

According to Driesch the fragment regulates itself toward the "whole" in the sense that it returns, at the beginning, to the structure from which the normal development of the whole also takes its start. As soon, therefore, as regulation is completed, the development of the fragments proceeds exactly like that in the normal. But such a regulation toward the whole appears possible, according to Driesch, only if the initial structure from which both courses of development start is a very simple one. Driesch reduces it, indeed, to a polar and bilaterally symmetrical orientation in the minutest parts. Driesch's real problem is that differentiation is possible on the basis of such a structure; from this point his theoretical views take their origin and herein lies one of the roots of his vitalism.

Roux, on the other hand, sets up a sharp distinction between primary (normal) and secondary (postgenerative) development. According to him the former proceeds in the frog's egg largely under self-differentiation of the four first blastomeres at least; the latter is of a more epigenetic character. This distinction rested upon the observations made in the pricking experiments, upon the self-differentiation of the surviving blastomere during early development, and upon the phenomena of postgeneration which manifested themselves in later development only. All the regulative potencies which the isolation experiment revealed in the germ fragments need not become active at all during the perfectly normal development. They would take part only in counteracting the effects of disturbances.

For the comprehension of normal development it is evidently of decisive significance to know which of the two theories is the true one; from the standpoint of purely theoretical interest in the phenomena of regulation, however, it is not so important. It is of theoretical interest that such a phenomenon exists at all, quite independent of the part it plays in normal development.

1938

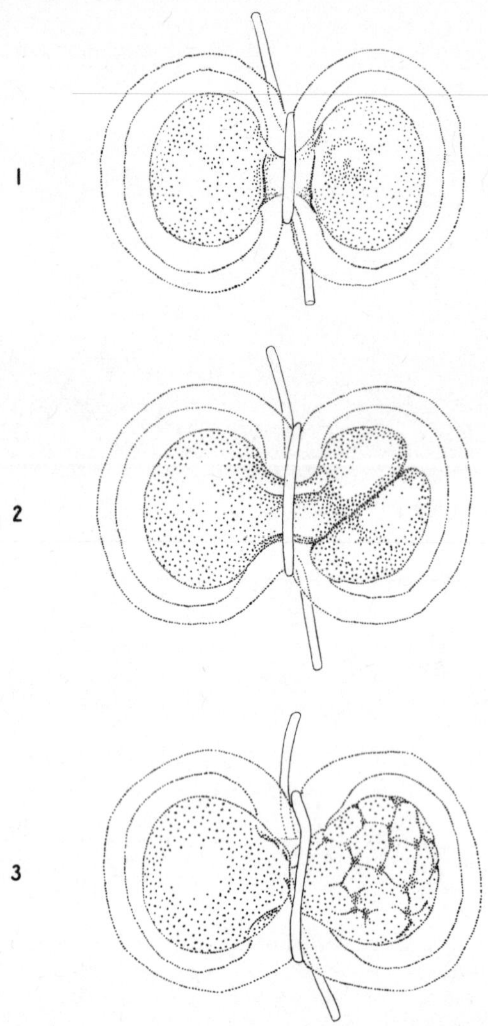

FIG. 1. Egg of the common newt, *Triton taeniatus;* constricted with a hair loop some time after fertilization. (After Spemann, 1919.)

FIG. 2. Egg of the newt in the two-celled stage. The egg nucleus has been retained, by constriction, in one half of the egg. (After Spemann, 1928.)

FIG. 3. The same operation as in Fig. 2, later. A descendant of the fertilized egg nucleus has passed the ligature into the other half of the egg. Both parts are separated by a furrow. (After Spemann, 1928.)

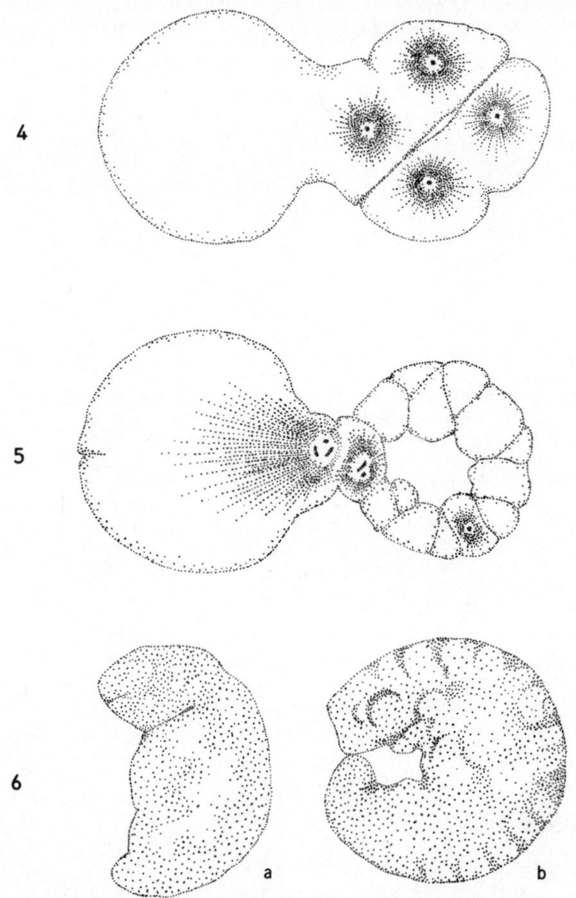

FIG. 4. The same, after moderate constriction in section. The half containing a nucleus has segmented twice; the nucleus which lies nearest to the plasmic bridge will send a descendant into the other half. (After Frankhauser, 1930a.)

FIG. 5. The same, after strong constriction, corresponding to Fig. 3. The nucleus which lies nearest to the plasmic bridge has sent a descendant into the other half; the latter is about to be cut off by a furrow. (After Frankhauser, 1930a.)

FIG. 6 *a* and *b*. Twins which have been produced by median constriction of the unsegmented egg. Embryo *a*, which received its nucleus later, is much younger than embryo *b*. The same phenomenon may be observed in Fig. 7 and Fig. 8. (After Spemann, 1928.)

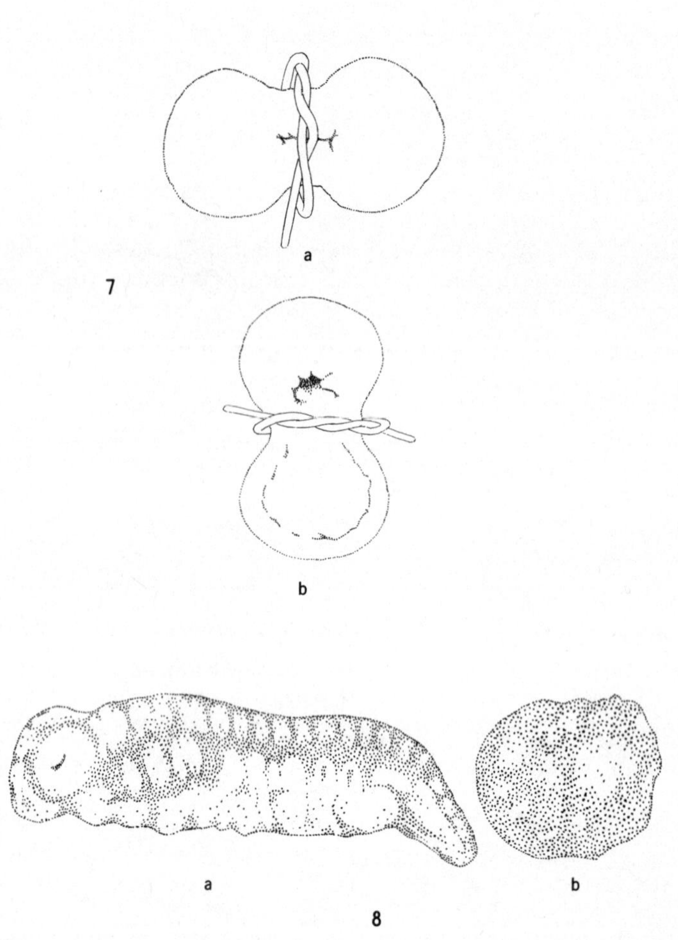

FIG. 7 *a* and *b*. Two germs of *Triton taeniatus* at the beginning of gastrulation. They had been weakly constricted in the two-cell stage, along the first furrow, which was either median (a) or frontal (b). (After Spemann, 1902 and 1903.)

FIG. 8 *a* and *b*. "Twins" resulting from frontal constriction of the fertilized egg. The dorsal half develops into a well-proportioned embryo; the ventral half forms a round piece without exterior differentiation. (After Spemann, unpublished.)

REFERENCES

Barfurth, D. (1893). Halbbildung oder Ganzbildung von halber Grösse? *Anat. Anz.,* 8:493–497.

Born, G. (1893). Über Druckverusche an Froscheiern. *Anat. Anz.,* 8.

Boveri, Th. (1910a). Über die Teilung centrifugierter Eier von Ascaris megalocephala. Festchr. f. Prof. Roux. II Teil. *Arch. f. Entw. Mech.,* 30:101–125.

Brachet, A. (1906). Recherches expérimentales sur l'oeuf nonsegmenté de Rana fusca. *Arch f. Entw. Mech.,* 22:325–341.

Driesch, H. (1892). *Idem,* IV. Experimentelle Veränderung des Typus der Furchung und ihre Folgen (Wirkung von Wärmezufuhr and Druck). *Ztschr. f. wiss. Zool.,* 55.

Endres, H. (1895). Über Anstich- und Schnürversuche an Eiern von Triton taeniatus. Schles. Ges. vaterl. Kult. 73. Jahresber.

Frankhauser, G. (1930a). Zytologische Untersuchungen an geschnürten triton-Eiern. 1. Die verzögerte Kernversorgung einer Hälfte nach hantelförmiger Einschnürung des Eies. *Roux's Arch.,* 122:116–139.

——— (1930b). Die Entwicklungspotenzen diploidkerniger Hälften des ungefurchten Tritoneies. *Roux's Arch.,* 122:672–735.

Harrison, R. G. (1918). "Experiments on the development of the fore limb of Amblystoma, a self-differentiating equipotential system." *Jour. Exp. Zool.,* 25:413–461.

Herlitzka, A. (1897). Sullo sviluppo di embrioni completi da blastomeri isolati di uova di tritone (Molge cristata). *Arch. f. Entw. Mech.,* 4:624–658.

Hertwig, O. (1893). Über den Wert der ersten Furchungszellen für die Organbildung des Embryo. *Arch. f. mikr. Anat. u. Entw. Mech.,* 42:662–806.

Lillie, Fr. R. (1929). "Embryonic segregation and its role in the life history." *Roux's Arch.,* 118:499–533.

Loeb, J. (1894). Über eine einfache Methode, zwei oder mehr zusammengewachsene Embryonen aus einem Ei hervorzubringen. *Pflügers Arch.,* 55:525–530.

McClendon, J. F. (1910). "The development of isolated blastomeres of the frog's egg." *Am. Jour. An.,* 10.

Morgan, T. H. (1895). "Half embryos and whole embryos from one of the first two blastomeres." *Anat. Anz.,* 10.

Schleip, W., und Penners, A. (1925). Über die Duplicitas cruciata bei den O. Schultzeschen Doppelbildungen von Rana fusca. *Verh. d. phys. med. Ges. zu Würzburg* (N. F.), 50.

——— (1926). Weitere Untersuchungen über die Entstechung der Schultze'schen Doppelbildungen beim braunen Frosch. *Verh. d. phys. med. Ges. zu Würzburg,* 51.

Schmidt, G. A. (1933). Schnürungs- und Durchschneidungsversuche am Anurenkeim. *Roux's Arch.,* 129:1–44.

Schütz, H. (1924). Schnüverusche an Tritoneiern vor Beginn der Furchung. Dissertation, unpublished (cf. Spemann, 1928).

Schultze, O. (1894). Die künstliche Erzeugung von Doppelbildungen bei Froschlarven mit Hilfe abnormer Gravitationswirkung. *Arch. f. Entw. Mech.,* 1:269–305.

Spemann, H. (1901). Entwicklungsphysiologische Studien am Tritonei I. *Arch. f. Entw. Mech.,* 12:224–264.

——— (1902). *Idem, II. Arch f. Entw. Mech.,* 15:448–534.

——— (1903). *Idem, III. Arch f. Entw. Mech.,* 16:551–631.

——— (1914). Über verzögerte Kernversorgung von Keimteilen. *Verh. d. D. Zool. Ges. Freiburg,* pp. 216–221.

——— (1919). Experimentelle Forschungen zum Determinations- und Individualitätsproblem. *Naturwissenschaften,* 7.

Spemann, H., und Falkenberg, H. (1919). Über asymmetrische Entwicklung und Situs inversus viscerum bei Zwillingen u. Doppelbildungen. *Arch. f. Entw. Mech.,* 45:371–422.

Vogt, W. (1927a). Über Hemmung der Formbildung an einer Halfte des Keims. (Nach Veruschen an Urodelen) Verh. Anat. Ges. 36 Vers. Kiel. *Anat. Anz.,* Erg. Heft, 63:126–139.

——— (1928a). Ablenkung der Symmetrie durch halbseitige Beschleunigung der Frühentwicklung (nach Versuchen an Pleurodelesund Axolotlkeimen). Verh. Anat. Ges. 37 Vers. Frankfurt a. M. *Anatol. Anz.,* Erg. Heft, 66:139–155.

——— (1928b). Mosaikcharakter und Regulation in der Frühentwicklung des Amphibieneies. Verh. d. D. Zool. Ges. 32 Vers. Munchen, pp. 26–70.

Wetzel, G. (1895). Über die Bedeutung der circulären Furche in der Entwicklung Furche der Schultze'schen Doppelbildungen von Rana fusca. *Arch. f. mikr. Anat. u. Entw. Mech.,* 46.

J. T. Bonner

Morphogenesis: An Essay on Growth

The slime mold *Dictyostelium discoides* provides an unusual opportunity for the experimental analysis of morphogenetic movements and differentiation. Bonner demonstrates, through simple experiments, that a chemical substance is responsible for the morphogenetic movements of the amoeboid cells. The chemical basis for the movement of the sluglike slime mold after aggregation is not known. Nor has a chemical basis been demonstrated for the culmination stage, during which the fruiting body (sporangium) is differentiated from the stalk and disk. Recent experiments with *Dictyostelium* prove that mutations may affect any stage of development in the life cycle. Some mutants remain in a myxameboid state; some form a slug but do not differentiate; some form several stalks and sporangia. The lack of a sexual system prevents a developmental genetic analysis of this system.

Much is known of the control of morphogenetic movements in the convergent *Acrasiales* or amoeboid slime molds. This particular group is of special interest to me, for I have been studying them for some years now. Because their nutrient requirements and culture conditions are known, and because their life cycle is so extremely short, lasting only four days, these organisms have been well suited for experimental studies. Especially so is the species *Dictyostelium discoideum,* for its fruiting structure has the most regular and simplest proportions of them all.[1]

The spores of *Dictyostelium* are small and capsule-shaped. Each one, when sown on moist agar, splits and hatches one unicellular amoeba. This amoeba soon begins to feed by engulfing bacteria and divides repeatedly by binary fission. In this vegetative stage of *Dictyostelium* the daughter cells are separate and free-swimming, wandering about quite independently of one

Reprinted from *Morphogenesis: An Essay on Growth* (Princeton University Press, 1952), pp. 173–183.

another. Their movement is a typical amoeboid or pseudopodial one, and this will be the first case we shall examine of coordinated morphogenetic movements involving amoeboid motion. If the amoebae reach a sufficient number (this critical number may be lowered if the food supply is depleted) then the aggregation starts. First a few and then finally all the amoebae start streaming in to central collection points (Fig. 1). Each amoeba does not make a bee-line to the center, but they come together to form streams, like tributaries collecting to form one large river.

The aggregated cell mass, containing anywhere from a few hundred to 100,000 cells, becomes elongate and will vary in length from 0.1 millimeter to 2 millimeters. Its size is entirely dependent on the number of amoebae that entered into the aggregation circle; there is no evidence of any growth, feeding, or cell division from the moment aggregation starts. This means that there is a fair likelihood that in this case growth and morphogenetic movements are completely separate. The growth is a random undirected process here, and all the molding and sculpturing is done by morphogenetic movements after the growth. The first such movement is that of aggregation, and now the sausage or bullet-shaped cell mass begins a second movement, that of migration. The whole mass glides about the agar at a speed comparable to the speed of an amoeba (roughly 2 mm per hour) giving off from its posterior end a slime streak which is in fact a sausage casing, produced by the sausage of cells, which collapses behind as the cell mass moves (Fig. 2d).[2] With the high powers of the microscope it is possible to see that the amoebae themselves are in active pseudopodial motion and somehow they exude the track and then walk on it, as the carpet of a Persian king is rolled before him. The sheath does not move, the cells move forward from within it, getting their traction on the sheath itself. The length of the migration period is very variable; it may either not exist at all, or it may extend, as we have been able to show recently, up to two weeks, if the environmental conditions are just right.

The third and last morphogenetic movement is somewhat more involved; it is the final fruiting or culmination where the migrating cell mass rights itself and shoots up into the air to form a delicate stalk supporting at its apex a round smooth spore mass (Fig. 2L). The anterior cells become the stalk cells, that is, they

as more prestalk cells climb to the apex and become stalk cells. The process is the reverse of a fountain; the cells pour up the outside to become trapped and solidified in the central core which is the stalk. In so doing the whole structure rises into the air until all the prestalk cells have been used up. Sometime in the beginning of the rising the prespore cells all became encapsulated into spores, so the final spore mass is resting at the apex.

Some while ago I became interested in the mechanism of the control of the morphogenetic movement of aggregation (Bonner, 1947). The early workers (Olive, 1902; Potts, 1902) had rather tacitly assumed that the process was by chemotaxis, by the orientation of the myxamoebae within a chemical gradient of some active substance, although there was no real evidence that this was so. More recently E. Runyon (1942) showed that if an attracting center was on one side of a semipermeable cellophane membrane, and the amoebae on the other, the amoebae would orient toward the center even though they were separated by the membrane. Runyon concluded from this ingenious experiment that the incoming amoebae must be guided by chemotaxis, but in fact his evidence was negligible for there are many things which may pass through a cellophane membrane besides chemical substances.

The one mechanism that was certainly impossible was that the amoebae agglutinated, for such forces must operate at close distances, and the thickness of cellophane would be far too great for such forces to be acting across it. A further proof of this came from a technique where the amoebae would aggregate in the bottom of a glass dish filled with water. If in such a preparation a central mass of cells was removed and placed quite a distance from the stream of cells that had been leading to it, three to five minutes afterward each cell in the stream separated from its neighbors and made a bee-line to the central mass at its new location. In this manner one can demonstrate that the amoebae will be attracted to masses that are half a millimeter away, which is equivalent to about sixty amoeba diameters.

In a number of ways, which I shall not enter into here, it was possible to show that any electric or magnetic force was most unlikely. Another possibility was that some sort of ray might guide the amoebae inward, as ships are guided into a harbor by a beacon. It was A. Gurwitsch, with his mitogenetic rays, who imagined that the control of all development was somehow achieved by these rays, although now the evidence of their existence is no longer considered adequate.[3] It was possible to show

that no radiation was involved in the aggregation of *Dictyostelium*. This was done by an experiment in which a thin glass shelf was suspended in water. A central mass of cells (i.e., a center) was placed on the upper side and separate amoebae on the underside (Fig. 3). After a time the separate amoebae all rounded the edge to the upper side and streamed in to the center. If the center had been a beacon emitting rays, the rays could either have penetrated the glass or not. If they had penetrated one would have expected the same result as with Runyon's cellophane; the amoebae would merely have gathered on the nearest point on the other side of the glass. If the rays had not passed through the glass, one would have expected the individual amoebae to be completely unaffected by the center. Since the amoebae were attracted around a sharp corner, and since no ray can bend around a corner, we can say that rays, mitogenetic or otherwise, do not play a part in the aggregation of *Dictyostelium*.

Another possibility was that some kind of Langmuirian film might guide the amoebae along the glass-water interface, an idea stimulated by an interesting experiment of Fauré-Fremiet and Wallich (1925) in which they showed, by placing talc on an interface, that there was a visible expanding of materials on the surface before the outward migration of amoebocytes of *Arenicola* (and of various types of cells in tissue culture). To test this hypothesis two submerged glass shelves were placed side by side, leaving a narrow gap between them. A center was placed on top of one of the plates and separate amoebae on top of the other. Again all the separate amoebae streamed to the edge of the shelf nearest the center and, since the amoebae could not bridge the gap, they formed a center right at the edge (Fig. 3). This proved that the attraction can occur across a region where there is no glass-water interface, and it is known that the amoebae need an interface for locomotion. The amoebae actually appear eager to cross the gap, for one can see them pawing the air with their pseudopods in a very frustrated manner, and if the two glass shelves are pushed together so that the gap is only one amoeba's length, the amoebae will immediately form a hanging bridge to join the center at last.

There are a number of other hypotheses that can be and have been tested, but among those only one showed positive evidence, that of chemotaxis. If aggregation takes place on the bottom of a dish, and the water above the amoebae flows over them gently, as though they were lying at the bottom of a brook, then only

the amoebae downstream are oriented toward the center. The amoebae upstream show no interest at all in the center. This must mean that the attracting agent is capable of diffusing, for it can be washed downstream. The only two diffusing agents are heat or chemical substances, and for a number of reasons which I shall not discuss here, heat seems unlikely, leaving a chemical substance as the only possibility.

The center, of course, does not really attract the amoebae in the sense of pulling them in. Presumably the chemical substance that is given off by the center is always more abundant near the center than farther away. Therefore, the front end of the amoebae approaching the center will be surrounded by more of the substance than will the hind end. In other words, all the substance accomplishes is to orient the amoebae by affecting the front end differentially from the hind end, the movement being solely that of the amoebae. The smell of carrion does not pull at the jackal, but it gives him a guide to where he may find the carcass, and the nearer he gets, the stronger the scent. The evidence is thus extremely good for the existence of a chemical influence in the aggregation of *Dictyostelium*. If there were some way of isolating the substance *in vitro* we would have a proof of its existence but so far all our attempts in this direction have failed.

In Edmund Spenser's *Faerie Queene* there is a witch named Acrasia, and like Circe, she attracts men and transforms them into beasts. It was decided that the chemical substance should be at least tentatively named, and considering its classical reference and the fact that *Dictyostelium* is a member of the *Acrasiales, acrasin* seemed entirely appropriate.

There are many aspects to the development of *Dictyostelium* that are pertinent to the subject matter of this book; in its aggregation we have seen an excellent instance of morphogenetic movement, and furthermore we seem to understand the mechanism whereby it is controlled. But there are other morphogenetic movements besides the aggregation movement, and to what extent may they also be interpreted in terms of gradients of acrasin? The first step was to find if acrasin was produced during the later stages of development, and it was possible not only to demonstrate its presence, but also to show, by two independent means, what was the ability of different parts of the cell mass to emit acrasin.[4]

Despite some earlier evidence to the contrary,[5] it now appears that the tip remains throughout development as the high point

of acrasin production, and so from this point of view it is conceivable that acrasin continues to be the guiding principle of movement. However, at some stages of migration there are no gradients of acrasin all along the main part of the migrating mass, with the exception of the region adjacent to the tip. Since the movement of the migrating cell mass is the same when there is an even emission along its axis, or when there is a gradient, it might be assumed that at least an external gradient of acrasin is not necessary for the directed movement of all parts of the cell mass. The same phenomenon is demonstrated in aggregation, where whirlpool aggregation patterns are sometimes obtained;[6] that is, instead of the cells coming in directly to one point they come in at an angle and make a central ring of amoebae with a hollow center. The cells in this ring go round and round following each other like a circle of elephants in the circus, and here again there appears to be directed movement without any overall acrasin gradient.

The argument then is that not all the morphogenetic movements of *Dictyostelium* are controlled by *external* acrasin gradients, gradients in the outside environment, but this does not mean that there might not be *internal* gradients that we cannot see or measure. It is quite conceivable, although treacherously hypothetical, that there may be some sort of gradient within each cell and that these line up with respect to one another much as Rashevsky (1938) imagines might be possible in his purely formal analysis of morphogenetic movements of animals. It is also possible that the contact, the adhesion between the cells, works in conjunction with acrasin gradients, so where the cells are not oriented by acrasin, they are guided by the movement tensions of the cells that surround them. The idea here then is that chemotaxis and contact guidance might both be simultaneously operating in *Dictyostelium*.

Even if the movements were entirely coordinated by gradients there would still be the problem (as there was with Child's metabolic gradients) of how the gradient was established in a particular configuration. There is no problem in explaining this for aggregation, because one needs only to postulate that each amoeba produces acrasin; and if more amoebae happen to be in one spot at the onset of aggregation, that spot will become the center, the high point in the acrasin gradient. But the problem is greater in migration, for there a uniform external gradient

consistently reverts to a high point at the apex and tapers off in the posterior direction. One might argue that there is an invisible external gradient imposed by aggregation that blossoms and becomes evident during the latter part of the migration period. The difficulty is that in one experiment[8] when vegetative amoebae were shaken in water, after a few minutes they stuck together in balls, and with twelve hours of shaking the balls were arrow-shaped, with a head and a tail end, the latter giving off a slime track. These also produced acrasin and somehow despite the shaking and despite the absence of normal aggregation, the cells became coordinated so that they had polarity and their morphogenetic movement (as exhibited by the continued production of slime track) was oriented in one direction only. So we have come back to a familiar barrier in our explanations; we need to know how the spatial configuration of acrasin distribution and of the polarity of movement were determined in the first place.

1952

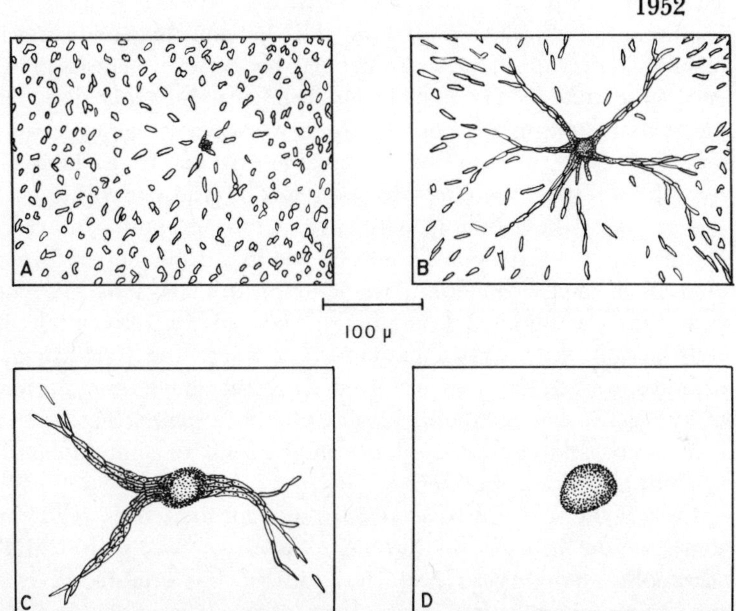

FIG. 1. A semi-diagrammatic representation of 4 stages of the aggregation of *Dictyostelium* taking place under water on the bottom of a glass dish. A, the beginning of aggregation showing the formation of a small center; B, C, successive stages of aggregation showing the thickening of the streams and the enlargement of the center; D, the final cell mass. (From Bonner, 1947.)

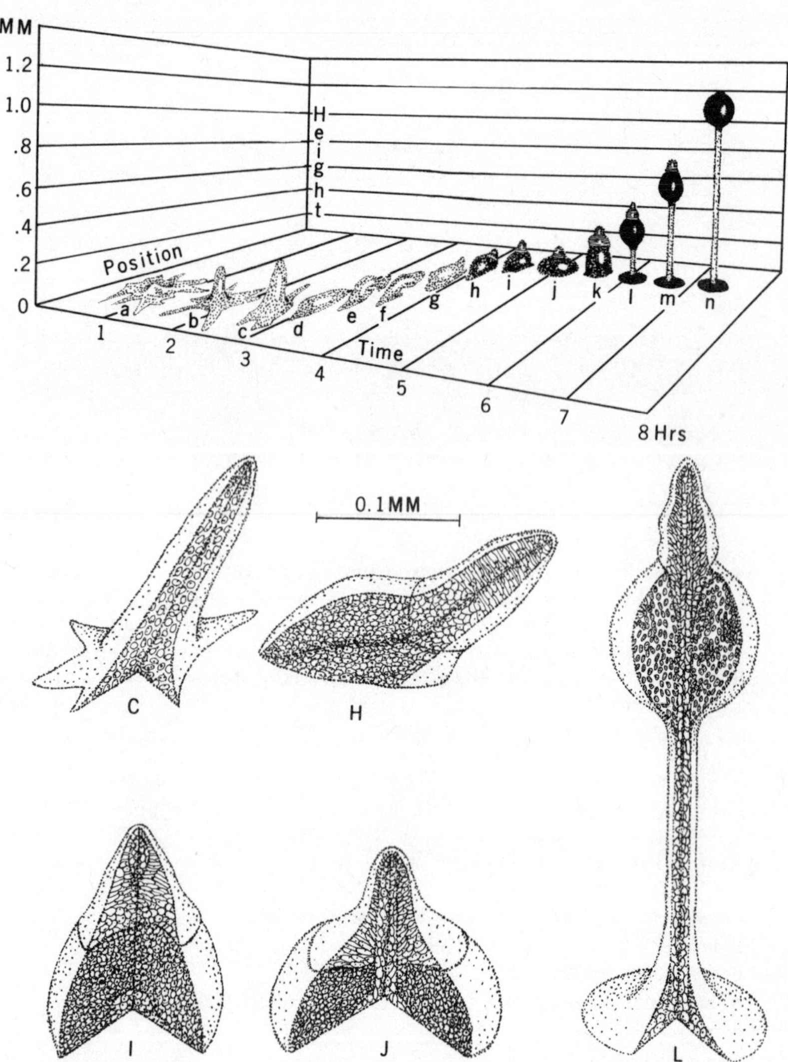

FIG. 2. Development in *Dictyostelium discoideum. Above:* The complete morphogenesis is represented in a three-dimensional graph. a–c, aggregation; d–h, migration; i–n, culmination. The presence of prespore cells is indicated by a heavy stippling, h–k; and the presence of true spores by solid black, l–n. *Below:* semi-diagrammatic drawings showing the cell structure at different stages. The letters indicate the corresponding stages given above. (From Bonner, 1944.)

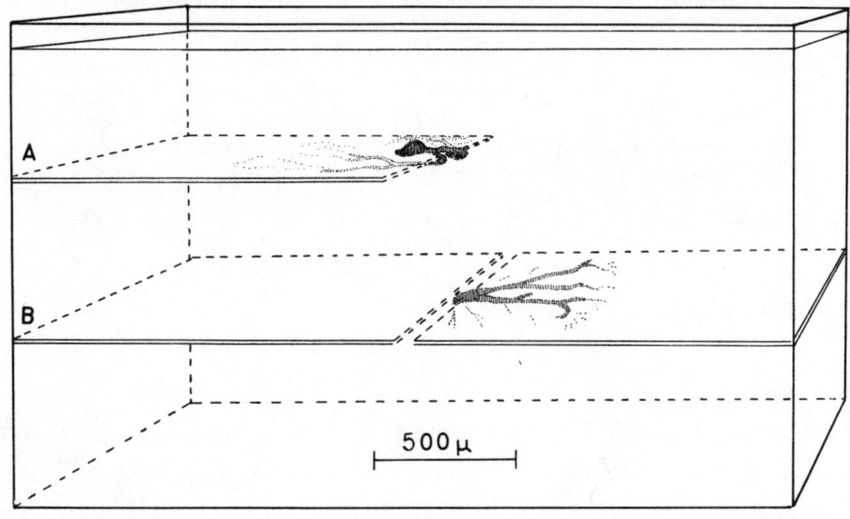

FIG. 3. A semi-diagrammatic representation of two experiments done on aggregation in *Dictyostelium* using coverslip shelves held under water. A, the myxamoebae previously at random under the coverslip are attracted around the edge to the center of the upper surface; B, the myxamoebae previously at random on the right-hand coverslip are attracted to the center on the left-hand coverslip, across the substratum gap. (Bonner, 1947.)

NOTES

1 For a description of the life cycle see Raper (1935, 1940a), Bonner (1944).

2 The migrating pseudoplasmodium will move toward light, and it has been shown recently that it is exceptionally sensitive to temperature gradients, migrating toward warmer regions (Bonner, Clarke, Neeley, and Slifkin, 1950).

3 See Hollaender (1936) for a general discussion of the problem.

4 Bonner (1949).

5 Bonner and Slifkin (1949).

6 Arndt (1937), Raper (1941).

7 Bonner (1950).

8 Bonner (1950).

REFERENCES

Arndt, A. (1937). Untersuchungen über *Dictyostelium mucoroides* Brefed. Roux' Arch. Entwickl., 136:681-744.

Bonner, J. T. (1944). "A descriptive study of the development of the slime mold *Dictyostelium discoideum*." *Amer. J. Bot.*, 31:175-182.

Bonner, J. T. (1947). "Evidence for the formation of cell aggregates by chemotaxis in the development of the slime mold *Dictyostelium discoideum.*" *J. Exp. Zool.,* 106:1-26.

Bonner, J. T. (1949). "The demonstration of acrasin in the later stages of the development of the slime mold *Dictyostelium discoideum.*" *J. Exp. Zool.,* 110:259-271.

Bonner, J. T. (1950). "Observations on polarity in the slime mold *Dictyostelium discoideum.*" *Biol. Bull.,* 99:143-151.

Bonner, J. T., W. W. Clarke, Jr., C. L. Neely, Jr., and M. K. Slifkin (1950). "The orientation to light and the extremely sensitive orientation to temperature gradients in the slime mold *Dictyostelium discoideum.*" *J. Cell. Comp. Physiol.,* 36:149-158.

Bonner, J. T., and M. K. Slifkin (1949). "A study of the control of differentiation: the proportions of stalk and spore cells in the slime mold *Dictyostelium discoideum.*" *Amer. J. Bot.,* 36:727-734.

Fauré-Fremiet, E., and R. Wallich (1925). "Un facteur physique de mouvement cellulaire pendant la culture des tissus *in vitro.*" *Compts. Rendu. Acad. Sci.,* 181:1096-1097.

Hollaender, A. (1936). "The problem of mitrogenetic rays." From B. M. Duggar's *Biological Effects on Radiation.* Vol. II, pp. 919-959.

Olive, E. W. (1902). Monograph of the *Acrasieae. Proc. Boston Soc. Nat. Hist.,* 30:451-513.

Potts, G. (1902). "Zur Physiologie des *Dictyostelium mucoroides.*" *Flora* (Jena), 91:281-347.

Raper, K. B. (1935). "*Dictyostelium discoideum,* a new species of slime mold from decaying forest leaves." *J. Agric. Res.,* 50:135-147.

Raper, K. B. (1940). "The communal nature of the fruiting process in the Acrasieae." *Amer. J. Bot.,* 27:436-448.

Raper, K. B. (1941). "Developmental patterns in simple slime molds." *Growth,* Symposium, 5:41-76.

Rashvesky, N. (1938). *Mathematical Biophysics.* (rev. ed.). Univ. of Chicago Press.

François Jacob and Jacques Monod

Structural and Regulatory Genes in the Biosynthesis of Proteins

Traditionally there have been two schools of thought on the mechanisms of differentiation. One view assumes that the cytoplasm plays a passive role, serving as a receptacle for instructions from the genes. This view leads to a paradox—how do cells with identical genotypes result in diverse tissue types? The second view assumes the presence of unknown cytoplasmic elements ("plasmagenes") or the existence of different states of the cytoplasm. These plasmatic features would modify the nucleus or themselves result in a changed cell type.

The genetic material has been viewed primarily as the source of information for enzyme and protein synthesis. It was Jacob and Monod who provided an experimental basis for a novel interpretation of the genetic basis of development. They proposed two kinds of genes—*structural genes,* which make proteins, and *regulatory genes,* which serve as "switches" to turn structural genes on or off. From the analysis of the β-galactosidase system in *Escherichia coli* Jacob and Monod located two types of regulatory genes. One type, the *i* gene, produces a product which reacts with the metabolite synthesized or utilized by the enzyme produced by a structural gene. The other gene, the *operator* or *o* gene, responds to the products of the *i* gene. The *o* region controls the simultaneous activities of several structural genes. The operator and its structural genes are all contiguous and form a line of several genes. Such a coordinated stimulation or repression of genes is characteristic of regulatory genes in bacteria. The group of genes coordinated by an operator is called an *operon.* The operon has not been found in higher cell systems but the search for such coordination has not been very intensive. The theory of the *operon* has given the first molecular insight into the possibility

From *Comptes Rendus de l'Academie des Sciences 249*:1282–1284, 1960, by permission of the publisher.

of a genetic control of development. The operon or other regulatory gene system has not yet been found in embryonic development but the belief that biological feedback systems exist is a dominant theme of contemporary embryologists.

It is acknowledged today that the structure of a protein is determined by a gene which provides the necessary information for its synthesis. It is also known that the expression of this information (the synthesis of the protein) is governed in many systems by certain specific substances which electively elicit or inhibit the synthesis of the protein. This is so, specifically, for inducible or repressible enzymes and also for lysogenic bacteria. At first, the mechanisms governing the expression of the "structural genes" in these different systems appear greatly dissimilar. Analysis of these systems, however, reveals certain remarkable analogies.

1. The synthesis of β-galactosidase and galactoside-permease in wild type *E. coli* is inducible by exogenous galactosides. Some constitutive mutants have been isolated in which these syntheses occur spontaneously. But these mutations have accrued in a cistron (i) which is independent in its expression from those cistrons which control the structures of the enzymes (z) and the permease (y). The inducible allele (i^+) is dominant over the constitutive (i^-) and the study of its expression in diploids indicates that it regulates the formation of a cytoplasmic repressor which inhibits the synthesis of the galactosidase and the permease, unless an exogenous inducer removes this inhibition.[1]

2. The formation of the sequence of enzymes responsible for the synthesis of tryptophan in the wild type *E. coli* is repressible by tryptophan.[2] Some non-repressible mutants have been isolated in which the repressor effect of tryptophan is abolished for all the enzymes of the sequence at the same time.

These mutations involve a "regulator" gene distinct from those which determine the capacity for synthesizing each of the enzymes individually. The repressible allele (R^+_{try}) of the "regulator" gene is dominant over the non-repressible allele (R^-_{try}). Its role, apparently, is the initiation (in the presence of tryptophan) of the synthesis of a repressor which inhibits the synthesis of each of the enzymes belonging to the sequence.[3]

3. In lysogenic *E. coli,* the expression of viral functions seems to be linked to the synthesis of specific proteins which are determined by the genes of the phage. These functions are inhibited

in prophages as readily as they are inhibited in phage genomes introduced by superinfection (immunity). In some merodiploids, heterozygous for the presence of prophage λ, the "immunity" character is dominant over "sensitivity." It is linked to a phage gene *c* whose expression in the cytoplasm of lysogenic bacteria seems to correspond to the presence of a repressor which should specifically inhibit one or more of those reactions (syntheses of proteins) which permit the phage genome to enter during the vegetative reproduction phase. Certain mutations of the gene suppress the capacity of the phage to lysogenize without altering its sensitivity to the repressor. These mutants behave as if they had lost the power to synthesize the repressor and they are comparable to the "constitutive" mutants of the galactoside-permease system and to the "derepressed" mutants of the anabolic system of tryptophan.[4] These observations justify, we believe, the conclusion that in the synthesis of many proteins a double genetic determinism exists, which activates two genes of distinct functions: the one (structural gene) responsible for the structure of the molecule, the other (regulatory gene) controlling the expression of the first through the action of a repressor. A particularly remarkable property of the regulatory genes, as we have defined them, is their pleiotropic effect: in the known examples, mutations which inactivate the regulatory genes affect simultaneously the synthesis of several proteins. This characteristic distinguishes the regulatory genes from the structural genes whose effects seem strictly limited to one protein, if not to one peptide chain. It seems, furthermore, that the expression of a regulatory gene does not imply the synthesis of a protein: the conversion of i^- *E. coli* cytoplasm into i^+ cytoplasm, that is, the synthesis of the repressor, is produced in the presence of substances (such as 5-methyltryptophan, *chloromycetin*) which block protein synthesis.[5] Similarly, in the lysogenic systems, the transition from the *non-immune* state to the *immune* state takes place in the presence of chloromycetin.[4] These results, whatever the biochemical interpretation is, also distinguish the regulatory genes from the structural genes, whose expression, by definition, implies the synthesis of a protein.

These observations pose the problem of the chemical nature of the repressor and of its site of action. We are not yet able to answer these questions. But it is important to emphasize a correlation which is undoubtedly significant. In the few instances known up till now, structural genes whose expression is subject

to the same regulatory gene (i.e., most probably subject to a unique repressor) are closely linked. The hypothesis can then be made that the action of the repressor depends upon a structure which is common to the group of genes whose expression it controls. Three possibilities may be considered: first, this common structure could be repeated in each of these genes (and in their cytoplasmic products); if this is the case, certain mutations could suppress the sensitivity to the repressor for the synthesis of each enzyme *separately*. These mutants would be *non-pleiotropic, dominant* and *constitutive*. Second, the structure would be a peptide chain common to the proteins of the group; it would be determined by an independent cistron, and it would be the site on which the repressor would have its effect. In this case, certain mutations of this cistron could suppress the sensitivity to the repressor for all the enzymes of the group simultaneously. These mutants would be *pleiotropic, dominant,* and *constitutive* and their effect would manifest itself in diploids in the *cis* as well as in the *trans* arrangement. Some other mutations of this cistron could result in the inactivation of the entire group of enzymes. They would be pleiotropic recessive mutations. These two hypotheses, however, do not take into account the "rule" of the grouping of genes whose expression is subjected to the same repressor; third, the expression of a group of genes would be linked to a unique structure, which would be sensitive to the repressor and would govern the activity of the group. A mutation occurring in this hypothetical structure, which one could call the "operator" of the group of genes which it would govern, could express itself by a loss of sensitivity to the repressor. Such mutants would appear to be pleiotropic, dominant and constitutive. Furthermore, their action should only be exerted upon structural genes placed in a *cis* position in relation to the "operator" in a diploid. This model would explain the grouping of structural genes subject to the same regulator, and, more generally, the correlation observed in the bacteria[6] between functional association and genetic linkage for sequential enzymatic systems. Notice, also, that simple mutations expressing themselves by the simultaneous loss of the capacity to synthesize several genetically linked proteins could be explained as effecting an "operator."[7]

1960

Translated by Patricia Girard.

NOTES

[1] A. B. Pardee, F. Jacob and J. Monod, *J. Mol. Biol.*, 1959 (in press).

[2] J. Monod and G. Cohen-Bazirre, *Comptes rendus,* 236, 1953, p. 530.

[3] G. N. Cohen and F. Jacob, *Comptes rendus,* 248, 1959, p. 3490.

[4] F. Jacob and A. M. Campbell, *Comptes rendus,* 248, 1959, p. 3219.

[5] A. B. Pardee and L. Prestidge (in press).

[6] M. Demerec, *Cold Spring Harb. Symp. Quant. Biol.*, 21, 1956, p. 113.

[7] This work has been beneficially supported by the Jarne Coffin Childs Memorial Fund and by the National Science Foundation.

R. S. Edgar and Millard Susman

Temperature Sensitive Mutants of T4D

Max Delbrück chose bacteriophage for biological study because they were molecules and their properties could be studied by physical and chemical means. The Hershey-Chase experiment proved that DNA was the genetical material which gave rise to new DNA and new protein in the formation of progeny virus. This phase of the study was then approached by three major investigations. Benzer analyzed the genetic fine structure of bacteriophage and related it to the Watson-Crick model of DNA. Several electron micrographers, notably T. Anderson and R. Horne, worked out the morphology of bacteriophage and their protein subunits. Finally, S. Cohen proved the existence of some dozen or more enzymes synthesized by the viral DNA during the early minutes of infection. These enzymes primarily affected the synthesis of new viral DNA. Using the information from all three lines of investigation, R. Edgar, his colleagues and his students were eager to find out how many genes were present in bacteriophage T4, which was the most widely used variety of bacteriophage.

Edgar's group used two different techniques to locate these genes. They found one class of mutations that was sensitive to high temperature but which could grow at normal temperature. These were called *conditional* mutations. A second class of conditional mutations (amber mutants, found by R. Epstein) could grow on one variety of *E. coli* host cell but could not grow on others. These two classes of mutations were believed to affect a *general* enzyme impairment rather than a single enzyme function. Tests of these mutants were carried out and about forty or more different genes were found involved. These were mapped and a surprising relation was found. Mutant genes affecting a similar function (e.g., DNA synthesis) were usually grouped together. It is not valid, yet, to conclude that an operon organization exists

Reprinted from *Research Reports 1961–1962 of the Division of Biology, California Institute of Technology,* Pasadena, California, pp. 145–148, by permission of the authors.

for developmental processes. Edgar's group is working, however, on the total genetic contribution to the developmental problems of how a bacteriophage is assembled from its subunit molecules.

Wild-type T4D forms plaques at an incubation temperature of 25°C with the same efficiency of plating as at an incubation temperature of 42°C. One can isolate "temperature-sensitive" mutants (*ts* mutants) which form plaques at 25°C but are unable to form plaques at 42°C. To date, 176 independent mutants have been isolated; 42 have been induced with nitrous acid, 33 with 2-aminopurine, 98 with 5-bromodeoxyuridine and one spontaneous mutant was also isolated. Mutant stocks can be grown at 25°C. The particles obtained are not more sensitive (in the majority of cases) to heat inactivation than the wild-type particles.

One would expect that mutants affected in different functions could complement each other. That is, upon mixed infection of one bacterium with two such different mutants, the cycle of growth of each could be completed at 42°C while each single mutant could not complete its cycle of growth. Such a test permits the division of the mutants into groups affecting the same function. On the basis of a "complementation spot test," the 176 mutants fall into about 50 complementation groups, or cistrons. Although all the possible pairwise combinations were not tested, all mutants fall into discrete classes giving negative tests within the group and only positive tests between groups. However, some groups of mutants behave in an exceptional fashion, giving some positive tests between members of the group. More refined complementation tests have been performed which consist of measuring the burst size of bacteria infected with various pairs of mutants, incubating them at the high temperature. Pairwise combinations of mutants give burst sizes either close to that of wild type (complementation), or a burst size of less than 20% of wild type (no complementation). The results of these experiments in general parallel the results of the spot tests.

Amber mutants are another class of mutants, studied by Epstein, which are widely scattered throughout the genome. Complementation spot tests have been performed between amber and *ts* mutants. A number of these tests were negative, indicating that some amber and *ts* mutants affect common functions. A map of the amber mutants (from data of Epstein) is included in the

figure, together with the correspondences between amber functional units and *ts* functional units.

To locate the mutants on the genome of T4D, one mutant from each cistron was selected for mapping. Standard crosses between *ts* mutants were performed and wild-type recombinants scored by plating at 42°C. On the basis of over 600 crosses, a map of the *ts* mutants has been constructed. The relationship of this map to the previously known map of T4D has been determined by crosses between *ts* mutants and plaque morphology mutants. The results are summarized in the figure. The *ts* mutations are widely distributed over the linkage structure. A number of crosses have been performed between mutants within the same cistron. In most cases the mutants fall in the same region of the genome; in general, the recombination values being less than 7%.

In one portion of the map the mutants used for mapping are rather densely but uniformly distributed. Utilizing data from crosses involving the mutants in this region, a mapping function, relating recombination values to map distance, was constructed, with a metric based on intervals averaging 6%. Because of the phenomenon of high negative interference, recombination values are not directly proportional to map distance. This mapping function permits the conversion of recombination values into map values, which are probably more closely related to physical distance along the chromosome. On the basis of this function, the map of T4D is greater than 800 map units in length. The distances presented in the figure are given as map values rather than recombination frequencies.

If bacteria are infected with a *ts* mutant and the complex incubated at a high temperature, the infection is abortive, since the gene or its product affected by the mutation is inoperative at that temperature. Studies on the nature of such abortive infection by different mutants throw light on the specific function of the affected gene. To this end, various observations have been made on abortive complexes: 1) The amount of DNA during the course of the infection was measured. A lack of increase would suggest that phage DNA synthesis is not initiated. 2) The complexes were examined visually in the electron microscope for characteristic cytological changes after infection (nuclear breakdown, etc.). 3) The optical density of the infected culture was followed for the occurrence of lysis. 4) If cells did not lyse, artificial lysates were made. Lysates were examined serologically for the presence of tail

fiber antigen. 5) Lysates were examined in the electron microscope for the presence or absence of morphological components of the phage.

The vast majority of the mutants are able to initiate DNA synthesis; and lysis of the cells occurs at the normal time with the production of phage specific material. However, only incomplete phage are produced. These mutants can be categorized in the following way. With one group of mutants morphologically normal looking phages are produced. These phages, however, are inactive. These mutants probably control the synthesis and assembly of tail fibers, since in this region are the previously studied host range and co-factor requirement mutations. Another group of mutants synthesize tail fiber protein and a normal number of tail cores to which are attached end-plates. However no heads, either full or empty, are manufactured. These mutations probably concern the synthesis and assembly of the head. These mutants are in the same region of the genome as the osmotic shock resistant mutant which has previously been studied. A third group of cistrons appear to affect the synthesis of tail components and their assembly since with these mutants a normal number of heads are synthesized, but the tail is either missing, unattached, or contracted.

Three mutants which affect DNA synthesis have been found. One mutant appears to be delayed in growth at the high temperature. However, the other two mutants appear to synthesize neither DNA, lysozyme, tail fiber protein, nor phage components. Nuclear breakdown does occur in these complexes.

The above preliminary results suggest that the genome of T4D shows a high degree of order, all mutations affecting similar functions being located in contiguous segments of the linkage structure.

In a given *ts* mutant a particular function is absent when growth takes place at a high temperature, but present at a low temperature. Information about the time during growth at which the gene acts can be obtained by transferring complexes from low to high or high to low temperature during the growth cycle. Some experiments of this type have been performed with a number of the mutants. The experiments support the following picture. Genes controlling the structure or assembly of phage particles act during terminal (tail functions) or near terminal (head functions) stages of growth (these may be designated late function genes). The genes

affecting DNA synthesis act early during growth ("early functions"). Late mutants complete their cycle of growth even at the high temperature although producing no infectious particles. The early mutants, on the other hand, appear to be arrested in growth at an early stage. The late functions depend upon the expression of the early functions for their initiation. If, at a time after the normal cycle of growth would have been completed, the complexes of early mutants are transferred to a lower temperature, they do resume the synthesis of phage.

1962

NOTE

Editor's Note: Prof. Edgar kindly permitted this slightly edited, but historically first note on his work with conditional lethals in bacteriophage T4D. The accompanying illustration, however, is a more accurate map which appears in the *Proceedings of the XI International Congress of Genetics,* the Hague, the Netherlands, Sept. 1963, vol. 2, edited by S. J. Geerts and published by Pergamon Press, 1965. For a detailed analysis of Edgar's contribution, see R. S. Edgar and R. H. Epstein, "Conditional Lethal Mutations in Bacteriophage T4" in the above *Proceedings,* pp. 1–15.

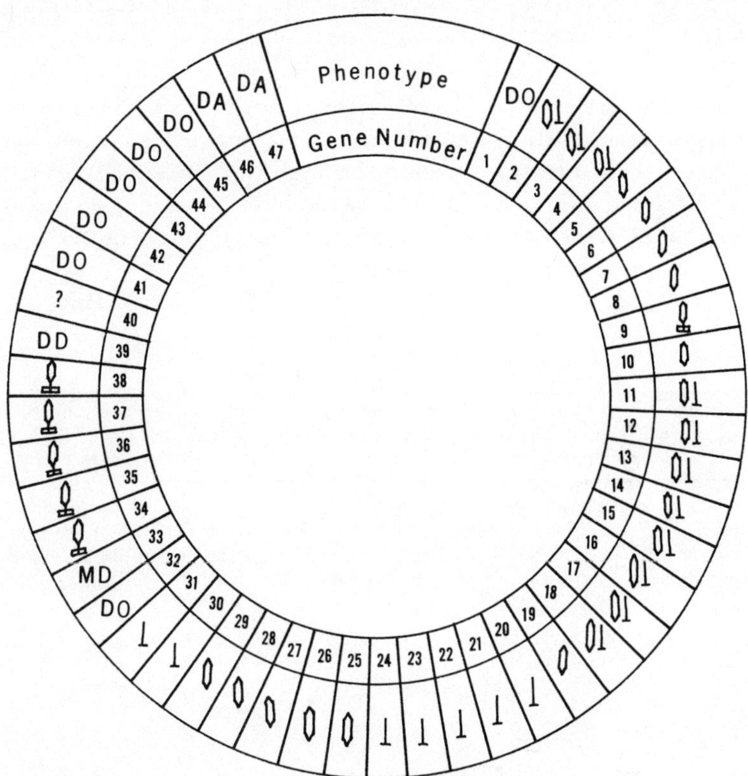

FIG. 1. A chart of the phenotypes of mutant genes of T4. A symbolic representation of the phenotype of each gene containing conditional lethals is given. The genes are presented in their order on the genetic map. The symbols represent the results of infections with mutants under restrictive conditions. DO(e.g. gene 1)—DNA synthesis does not occur. Late functions, e.g. the synthesis of lysozyme or of phage components, does not occur. The bacterial nucleus, however, is degraded and the synthesis of some early enzymes occurs. DA(e.g. gene 46)—DNA synthesis is initiated, then ceases after a short time. Late functions are expressed. DD (gene 39)—DNA synthesis is delayed for about 20 min., then proceeds normally; late functions show a corresponding delay. MD(e.g. gene 33)—DNA synthesis is normal but late functions are not expressed. *Heads and tails* (e.g. gene 2)—Unassembled head membranes and tail cores with attached end plates are formed in normal amounts. *Heads* (e.g. gene 5)—Only head membranes are formed. *Tail cores only* (e.g. gene 20)—Only tail cores with attached end plates are formed in normal amounts, heads are absent. A *phage particle* (e.g. gene 25)—Phage particles are present in normal amounts, but are inactive and, in some cases at least, lack tail fibers.

PART IV ○○○

Molecular Biology

MOLECULAR BIOLOGY DIFFERS FROM biochemistry in point of view rather than in technique or subject matter. In biochemistry, the chemical analysis of cellular phenomena, the cells themselves, with their internal organization, have usually been considered expendable. The recognition that structure and function are significant within the cell led to the new field of molecular biology. In this view, structure was important because of the size of the molecules. Macromolecules, such as DNA, RNA, and many of the proteins, are among the most significant metabolic components of the cell.

The instruments for studying these macromolecules include most of the tools used by biochemists, such as the centrifuge, chromatography, cell fractionation, radioactive labeling, and the various devices for detecting radioactivity. The *centrifuge* separates fragments of cells by their molecular weight. The heavier particles form layers at the bottom of a centrifuge tube and the lighter particles form layers on the topmost portion of the tube. The layering is caused by enormous centrifugal forces exerted on molecules of different molecular weights. *Chromatography* uses paper, cellulose, resins, or clay on which different molecules may assort themselves according to the rapidity with which they are carried by liquid solvents. The various layers or streaks of molecules can be removed and the molecules may be further purified. Chromatography can be used to separate different proteins from one another. This technique is sometimes augmented by passing an electric current through the solvent, thus using the electrical charges of the molecules as well as their chemical characteristics for the chromatogram. In *cell fractiona-*

tion, the processes that dismember cells into their fragments are carefully chosen so that the functions of the cell's organelles are still retained.

In addition to these techniques, molecular biologists rely on the *electron microscope* for observation of detailed structure of cell organelles. Electron microscopes now provide remarkable detail of virus structure, microbial DNA, cell membranes, and the internal structure of cell organelles. Another modern technique is *radioautography*. Compounds containing radioisotopes are fed to bacterial or other organisms and the labeled macromolecules assembled from the isotope can be studied by radioautographs superimposed on optical or electron micrographs.

The main feature of molecular biology has been the union of genetics, biochemistry, physics, anatomy, microbiology, and other scientific disciplines. The goal of this science is the eventual interpretation of the living state as the physical and chemical coordination of macromolecules.

In the 1950's molecular biology showed convincingly that DNA is the genetic material of most living things and that the Watson-Crick model of its structure provides the basis for gene replication. The molecular basis of mutation was demonstrated by use of base analogues, compounds similar to the nucleotides found in DNA. The sites of mutations in genetic maps of viruses and bacteria corresponded grossly to the structure of the DNA molecule. The "dogma" that DNA makes RNA and that RNA makes protein was shown to be more than a slogan of confidence in molecular biology. Many contributions were made, especially by M. Hoagland, to the artificial synthesis of proteins from the isolated components of the cell. The amino acids must be activated with an energy-rich compound; they are then combined with transfer RNA molecules. The genetic information must be transcribed from DNA to messenger RNA or it may be artificially synthesized. Ribosomes are also required for this system before the amino acids can be connected by an alignment of the transfer RNA molecules on the messenger RNA. The ribosomes bring about the protein synthesis from these other components.

In the 1960's molecular biologists demonstrated the existence of a universal genetic code. Nirenberg and Matthaei at the National Institutes of Health and Ochoa at NYU Medical School worked out the various nucleotide coding combinations (codons) for the amino acids. A language with four letters in its alphabet

(DNA or RNA) was translated into a language with twenty letters in its alphabet (protein). The technique used by Nirenberg and Matthaei is comparable to a nursery school syllabary. Messenger RNA molecules are constructed containing only one "letter." Thus the message "U.U.U.U.U. . . ." (poly-uracil) can be read as a sequence of codons containing only "U." In the Hoagland system for protein synthesis, the resulting "protein" is poly-phenylalanine: "phe.phe.phe.phe. . . ." By making simple words (e.g., "AUAUAUAU . . .") the various codons were deciphered. Each codon consists of three nucleotides; for this reason it is called a *triplet code.* Thus "UUU" is the codon for *one* amino acid molecule, phenylalanine.

The second major contribution of molecular biology in the early 1960's was the discovery of regulatory genes in bacteria. These genes control the activity of structural genes. Structural genes carry the specificity for enzymes while regulatory genes produce products which can switch the structural genes on or off. This regulatory function is associated with a biochemical feedback mechanism that prevents cells from producing too much or too little enzyme. The application of the theory of regulatory genes (known as "operons") to development is actively in progress. So too is the study of protein evolution through specific molecules (e.g., hemoglobin) found in a variety of different organisms.

The future of molecular biology cannot be predicted because so many novel discoveries have been made about life in the past ten years. If life has existed on earth for more than two billion (2×10^9) years, then we would be unduly optimistic to believe that its complexities will be unraveled in less than fifty years. There will be much knowledge and experimentation for the twenty-first century. One direction for the future is the molecular biology of nerve cells. Is memory stored by informational macromolecues (RNA) or by electric energy retained in neurone circuits?

Another direction is the artificial synthesis of enzymes. Determination of the structure of proteins was achieved by Kendrew and Perutz, who used X-ray diffraction techniques. The three-dimensional organization of the sequence of amino acids now permits biophysicists to test the function of these molecules. Their results may answer the question, How do enzymes work? This will be a major area of contemporary studies in molecular

biology. Once the three-dimensional structure of proteins can be related to their function, computers will be used to design proteins. A knowledge of the genetic code, in turn, will predict the DNA nucleotide sequence for coding such proteins. If biochemists devise techniques for synthesizing DNA according to desired sequences, then a new and imaginative branch of molecular biology will emerge—the design of artificial enzymes to carry out new or existing functions (e.g., digest cellulose; fix atmospheric gases into carbohydrates; synthesize any organic molecules). It will be possible to create new life forms using the predicted efficiencies of synthetic genes rather than haphazard collections of what natural selection has produced.

This may, at present, sound fanciful, but then who would have predicted in 1900 that the chemical and physical basis of heredity would be solved by 1966?

Linus Pauling, Harvey A. Itano, S. J. Singer, and Ibert C. Wells

Sickle Cell Anemia, a Molecular Disease

Linus Pauling has the rare distinction of having received two Nobel prizes—one for Chemistry and one for Peace. Yet two major contributions have been made by Pauling to biology. One of these is the alpha helix, which accounts for the stability and much of the linearity of protein molecules. The second is the recognition that genetic diseases in man may be a consequence of abnormal proteins formed by mutant genes. Pauling called the sickle-cell trait, which is inherited, a "molecular disease." The recognition that pathology can be extended from the organism, the organ, the tissue, and the cell to a specific molecule within that cell is a remarkable example of the progression of technique in pathology in this last century. No one knows how many diseases in man may be designated molecular diseases. Some forms of diabetes and such diseases as cystic fibrosis, muscular dystrophy, color blindness, hemophilia, and phenylketonuric idiocy are examples of known hereditary diseases which are likely to have defective proteins. Pauling's discovery actually confirms a theory proposed early in the 1900's by Sir Archibald Garrod. Garrod had examined several inherited diseases and concluded that they were "inborn errors of metabolism." Pauling's analysis of the sickle-cell trait shows how such an error of metabolism can be traced to a specific molecule.

The erythrocytes of certain individuals possess the capacity to undergo reversible changes in shape in response to changes in the partial pressure of oxygen. When the oxygen pressure is lowered, these cells change their forms from the normal biconcave disk to crescent, holly wreath, and other forms. This process is known

Reprinted from *Science 110*:543–548, 1949, by permission of the publisher and the authors.

as sickling. About 8 percent of American Negroes possess this characteristic; usually they exhibit no pathological consequences ascribable to it. These people are said to have sicklemia, or sickle cell trait. However, about 1 in 40 (4) of these individuals whose cells are capable of sickling suffer from a severe chronic anemia resulting from excessive destruction of their erythrocytes; the term sickle cell anemia is applied to their condition.

The main observable difference between the erythrocytes of sickle cell trait and sickle cell anemia has been that a considerably greater reduction in the partial pressure of oxygen is required for a major fraction of the trait cells to sickle than for the anemia cells (11). Tests *in vivo* have demonstrated that between 30 and 60 percent of the erythrocytes in the venous circulation of sickle cell anemic individuals, but less than 1 percent of those in the venous circulation of sicklemic individuals, are normally sickled. Experiments *in vitro* indicate that under sufficiently low oxygen pressure, however, all the cells of both types assume the sickled form.

The evidence available at the time that our investigation was begun indicated that the process of sickling might be intimately associated with the state and the nature of the hemoglobin within the erythrocyte. Sickle cell erythrocytes in which the hemoglobin is combined with oxygen or carbon monoxide have the biconcave disk contour and are indistinguishable in that form from normal erythrocytes. In this condition they are termed promeniscocytes. The hemoglobin appears to be uniformly distributed and randomly oriented within normal cells and promeniscocytes, and no birefringence is observed. Both types of cells are very flexible. If the oxygen or carbon monoxide is removed, however, transforming the hemoglobin to the uncombined state, the promeniscocytes undergo sickling. The hemoglobin within the sickled cells appears to aggregate into one or more foci, and the cell membranes collapse. The cells become birefringent (11) and quite rigid. The addition of oxygen or carbon monoxide to these cells reverses these phenomena. Thus the physical effects just described depend on the state of combination of the hemoglobin, and only secondarily, if at all, on the cell membrane. This conclusion is supported by the observation that sickled cells when lysed with water produce discoidal, rather than sickle-shaped, ghosts (10).

It was decided, therefore, to examine the physical and chemical properties of the hemoglobins of individuals with sicklemia

and sickle cell anemia, and to compare them with the hemoglobin of normal individuals to determine whether any significant differences might be observed.

EXPERIMENTAL METHODS

The experimental work reported in this paper deals largely with an electrophoretic study of these hemoglobins. In the first phase of the investigation, which concerned the comparison of normal and sickle cell anemia hemoglobins, three types of experiments were performed: 1) with carbonmonoxyhemoglobins; 2) with uncombined ferrohemoglobins in the presence of dithionite ion, to prevent oxidation to methemoglobins; and 3) with carbonmonoxyhemoglobins in the presence of dithionite ion. The experiments of type 3 were performed and compared with those of type 1 in order to ascertain whether the dithionite ion itself causes any specific electrophoretic effect.

Samples of blood were obtained from sickle cell anemic individuals who had not been transfused within three months prior to the time of sampling. Stroma-free concentrated solutions of human adult hemoglobin were prepared by the method used by Drabkin (3). These solutions were diluted just before use with the appropriate buffer until the hemoglobin concentrations were close to 0.5 grams per 100 milliliters, and then were dialyzed against large volumes of these buffers for 12 to 24 hours at 4° C. The buffers for the experiments of types 2 and 3 were prepared by adding 300 ml of 0.1 ionic strength sodium dithionite solution to 3.5 liters of 0.1 ionic strength buffer. About 100 ml of 0.1 molar NaOH was then added to bring the pH of the buffer back to its original value. Ferrohemoglobin solutions were prepared by diluting the concentrated solutions with this dithionite-containing buffer and dialyzing against it under a nitrogen atmosphere. The hemoglobin solutions for the experiments of type 3 were made up similarly, except that they were saturated with carbon monoxide after dilution and were dialyzed under a carbon monoxide atmosphere. The dialysis bags were kept in continuous motion in the buffers by means of a stirrer with a mercury seal to prevent the escape of the nitrogen and carbon monoxide gases.

The experiments were carried out in the modified Tiselius electrophoresis apparatus described by Swingle (14). Potential gradients of 4.8 to 8.4 volts per centimeter were employed, and the duration of the runs varied from 6 to 20 hours. The pH

values of the buffers were measured after dialysis on samples which had come to room temperature.

RESULTS

The results indicate that a significant difference exists between the electrophoretic mobilities of hemoglobin derived from erythrocytes of normal individuals and from those of sickle cell anemic individuals. The two types of hemoglobin are particularly easily distinguished as the carbonmonoxy compounds at pH 6.9 in phosphate buffer of 0.1 ionic strength. In this buffer the sickle cell anemia carbonmonoxyhemoglobin moves as a positive ion, while the normal compound moves as a negative ion, and there is no detectable amount of one type present in the other.[1] The hemoglobin derived from erythrocytes of individuals with sicklemia, however, appears to be a mixture of normal hemoglobin and sickle cell anemia hemoglobin in roughly equal proportions. Up to the present time the hemoglobins of 15 persons with sickle cell anemia, 8 persons with sicklemia, and 7 normal adults have been examined. The hemoglobins of normal adult white and Negro individuals were found to be indistinguishable.

The mobility data obtained in phosphate buffers of 0.1 ionic strength and various values of pH are summarized in Figs. 1 and 2.[2]

TABLE I

ISOELECTRIC POINTS IN PHOSPHATE BUFFER, $\mu = 0.1$

Compound	Normal	Sickle cell anemia	Difference
Carbonmonoxyhemoglobin	6.87	7.09	0.22
Ferrohemoglobin	6.87	7.09	0.22

The isoelectric points are listed in Table I. These results prove that the electrophoretic difference between normal hemoglobin and sickle cell anemia hemoglobin exists in both ferrohemoglobin and carbonmonoxyhemoglobin. We have also performed several experiments in a buffer of 0.1 ionic strength and pH 6.52 containing 0.08M NaCl, 0.02M sodium cacodylate, and 0.0083M cacodylic acid. In this buffer the average mobility of sickle cell anemia carbonmonoxyhemoglobin is 2.63×10^{-5}, and that of normal carbonmonoxyhemoglobin is 2.23×10^{-5} cm/sec per

volt/cm.[3] These experiments with a buffer quite different from phosphate buffer demonstrate that the difference between the hemoglobins is essentially independent of the buffer ions.

Typical Longsworth scanning diagrams of experiments with normal, sickle cell anemia, and sicklemia carbonmonoxyhemoglobins, and with a mixture of the first two compounds, all in phosphate buffer of pH 6.90 and ionic strength 0.1, are reproduced in Fig. 3. It is apparent from this figure that the sicklemia material contains less than 50 percent of the anemia component. In order to determine this quantity accurately some experiments at a total protein concentration of 1 percent were performed with known mixtures of sickle cell anemia and normal carbonmonoxyhemoglobins in the cacodylate–sodium chloride buffer of 0.1 ionic strength and pH 6.52 described above. This buffer was chosen in order to minimize the anomalous electrophoretic effects observed in phosphate buffers (7). Since the two hemoglobins were incompletely resolved after 15 hours of electrophoresis under a potential gradient of 2.79 volts/cm, the method of Tiselius and Kabat (16) was employed to allocate the areas under the peaks in the electrophoresis diagrams to the two components. In Fig. 4 there is plotted the percent of the anemia component calculated from the areas so obtained against the percent of that component in the known mixtures. Similar experiments were performed with a solution in which the hemoglobins of 5 sicklemic individuals were pooled. The relative concentrations of the two hemoglobins were calculated from the electrophoresis diagrams, and the actual proportions were then determined from the plot of Fig. 4. A value of 39 percent for the amount of the sickle cell anemia component in the sicklemia hemoglobin was arrived at in this manner. From the experiments we have performed thus far it appears that this value does not vary greatly from one sicklemic individual to another, but a more extensive study of this point is required.

Up to this stage we have assumed that one of the two components of sicklemia hemoglobin is identical with sickle cell anemia hemoglobin and the other is identical with the normal compound. Aside from the genetic evidence which makes this assumption very probable (see the discussion section), electrophoresis experiments afford direct evidence that the assumption is valid. The experiments on the pooled sicklemia carbonmonoxyhemoglobin and the mixture containing 40 percent sickle cell anemia carbonmonoxyhemoglobin and 60 percent normal carbon-

monoxyhemoglobin in the cacodylate–sodium chloride buffer described above were compared, and it was found that the mobilities of the respective components were essentially identical.[4] Furthermore, we have performed experiments in which normal hemoglobin was added to a sicklemia preparation and the mixture was then subjected to electrophoretic analysis. Upon examining the Longsworth scanning diagrams we found that the area under the peak corresponding to the normal component had increased by the amount expected, and that no indication of a new component could be discerned. Similar experiments on mixtures of sickle cell anemia hemoglobin and sicklemia preparations yielded similar results. These sensitive tests reveal that, at least electrophoretically, the two components in sicklemia hemoglobin are identifiable with sickle cell anemia hemoglobin and normal hemoglobin.

DISCUSSION

1) *On the Nature of the Difference between Sickle Cell Anemia Hemoglobin and Normal Hemoglobin:* Having found that the electrophoretic mobilities of sickle cell anemia hemoglobin and normal hemoglobin differ, we are left with the considerable problem of locating the cause of the difference. It is impossible to ascribe the difference to dissimilarities in the particle weights or shapes of the two hemoglobins in solution: a purely frictional effect would cause one species to move more slowly than the other throughout the entire pH range and would not produce a shift in the isoelectric point. Moreover, preliminary velocity ultracentrifuge[5] and free diffusion measurements indicate that the two hemoglobins have the same sedimentation and diffusion constants.

The most plausible hypothesis is that there is a difference in the number or kind of ionizable groups in the two hemoglobins. Let us assume that the only groups capable of forming ions which are present in carbonmonoxyhemoglobin are the carboxyl groups in the heme, and the carboxyl, imidazole, amino, phenolic hydroxyl, and guanidino groups in the globin. The number of ions nonspecifically absorbed on the two proteins should be the same for the two hemoglobins under comparable conditions, and they may be neglected for our purposes. Our experiments indicate that the net number of positive charges (the total number of cationic groups minus the number of anionic groups) is greater for sickle

cell anemia hemoglobin than for normal hemoglobin in the pH region near their isoelectric points.

According to titration data obtained by us, the acid-base titration curve of normal human carbonmonoxyhemoglobin is nearly linear in the neighborhood of the isoelectric point of the protein, and a change of one pH unit in the hemoglobin solution in this region is associated with a change in net charge on the hemoglobin molecule of about 13 charges per molecule. The same value was obtained by German and Wyman (5) with horse oxyhemoglobin. The difference in isoelectric points of the two hemoglobins under the conditions of our experiments is 0.23 for ferrohemoglobin and 0.22 for the carbonmonoxy compound. This difference corresponds to about 3 charges per molecule. With consideration of our experimental error, sickle cell anemia hemoglobin therefore has 2–4 more net positive charges per molecule than normal hemoglobin.

Studies have been initiated to elucidate the nature of this charge difference more precisely. Samples of porphyrin dimethyl esters have been prepared from normal hemoglobin and sickle cell anemia hemoglobin. These samples were shown to be identical by their X-ray powder photographs and by identity of their melting points and mixed melting point. A sample made from sicklemia hemoglobin was also found to have the same melting point. It is accordingly probable that normal and sickle cell anemia hemoglobin have different globins. Titration studies and amino acid analyses on the hemoglobins are also in progress.

2) *On the Nature of the Sickling Process:* In the introductory paragraphs we outlined the evidence which suggested that the hemoglobins in sickle cell anemia and sicklemia erythrocytes might be responsible for the sickling process. The fact that the hemoglobins in these cells have now been found to be different from that present in normal red blood cells makes it appear very probable that this is indeed so.

We can picture the mechanism of the sickling process in the following way. It is likely that it is the globins rather than the hemes of the two hemoglobins that are different. Let us propose that there is a surface region on the globin of the sickle cell anemia hemoglobin molecule which is absent in the normal molecule and which has a configuration complementary to a different region of the surface of the hemoglobin molecule. This situation would be somewhat analogous to that which very prob-

ably exists in antigen-antibody reactions (9). The fact that sickling occurs only when the partial pressures of oxygen and carbon monoxide are low suggests that one of these sites is very near to the iron atom of one or more of the hemes, and that when the iron atom is combined with either one of these gases, the complementariness of the two structures is considerably diminished. Under the appropriate conditions, then, the sickle cell anemia hemoglobin molecules might be capable of interacting with one another at these sites sufficiently to cause at least a partial alignment of the molecules within the cell, resulting in the erythrocyte's becoming birefringent, and the cell membrane's being distorted to accommodate the now relatively rigid structures within its confines. The addition of oxygen or carbon monoxide to the cell might reverse these effects by disrupting some of the weak bonds between hemoglobin molecules in favor of the bonds formed between gas molecules and iron atoms of the hemes.

Since all sicklemia erythrocytes behave more or less similarly, and all sickle at a sufficiently low oxygen pressure (11), it appears quite certain that normal hemoglobin and sickle cell anemia hemoglobin coexist within each sicklemia cell; otherwise there would be a mixture of normal and sickle cell anemia erythrocytes in sicklemia blood. We might expect that the normal hemoglobin molecules, lacking at least one type of complementary site present on the sickle cell anemia molecules, and so being incapable of entering into the chains or three-dimensional frameworks formed by the latter, would interfere with the alignment of these molecules within the sicklemia erythrocyte. Lower oxygen pressures, freeing more of the complementary sites near the hemes, might be required before sufficiently large aggregates of sickle cell anemia hemoglobin molecules could form to cause sickling of the erythrocytes.

This is in accord with the observations of Sherman (11), which were mentioned in the introduction, that a large proportion of erythrocytes in the venous circulation of persons with sickle cell anemia are sickled, but that very few have assumed the sickle forms in the venous circulation of individuals with sicklemia. Presumably, then, the sickled cells in the blood of persons with sickle cell anemia cause thromboses, and their increased fragility exposes them to the action of reticulo-endothelial cells which break them down, resulting in the anemia (1).

It appears, therefore, that while some of the details of this

picture of the sickling process are as yet conjectural, the proposed mechanism is consistent with experimental observations at hand and offers a chemical and physical basis for many of them. Furthermore, if it is correct, it supplies a direct link between the existence of "defective" hemoglobin molecules and the pathological consequences of sickle cell disease.

3) *On the Genetics of Sickle Cell Disease:* A genetic basis for the capacity of erythrocytes to sickle was recognized early in the study of this disease (4). Taliaferro and Huck (15) suggested that a single dominant gene was involved, but the distinction between sicklemia and sickle cell anemia was not clearly understood at the time. The literature contains conflicting statements concerning the nature of the genetic mechanisms involved, but recently Neel (8) has reported an investigation which strongly indicates that the gene responsible for the sickling characteristic is in heterozygous condition in individuals with sicklemia, and homozygous in those with sickle cell anemia.

Our results had caused us to draw this inference before Neel's paper was published. The existence of normal hemoglobin and sickle cell anemia hemoglobin in roughly equal proportions in sicklemia hemoglobin preparations is obviously in complete accord with this hypothesis. In fact, if the mechanism proposed above to account for the sickling process is correct, we can identify the gene responsible for the sickling process with one of an alternative pair of alleles capable through some series of reactions of introducing the modification into the hemoglobin molecule that distinguishes sickle cell anemia hemoglobin from the normal protein.

The results of our investigation are compatible with a direct quantitative effect of this gene pair; in the chromosomes of a single nucleus of a normal adult somatic cell there is a complete absence of the sickle cell gene, while two doses of its allele are present; in the sicklemia somatic cell there exists one dose of each allele; and in the sickle cell anemia somatic cell there are two doses of the sickle cell gene, and a complete absence of its normal allele. Correspondingly, the erythrocytes of these individuals contain 100 percent normal hemoglobin, 40 percent sickle cell anemia hemoglobin and 60 percent normal hemoglobin, and 100 percent sickle cell anemia hemoglobin, respectively. This investigation reveals, therefore, a clear case of a change produced in a protein

molecule by an allelic change in a single gene involved in synthesis.

The fact that sicklemia erythrocytes contain the two hemoglobins in the ratio 40:60 rather than 50:50 might be accounted for by a number of hypothetical schemes. For example, the two genes might compete for a common substrate in the synthesis of two different enzymes essential to the production of the two different hemoglobins. In this reaction, the sickle cell gene would be less efficient than its normal allele. Or, competition for a common substrate might occur at some later stage in the series of reactions leading to the synthesis of the two hemoglobins. Mechanisms of this sort are discussed in more elaborate detail by Stern (13).

The results obtained in the present study suggest that the erythrocytes of other hereditary hemolytic anemias be examined for the presence of abnormal hemoglobins. This we propose to do.

Based on a paper presented at the meeting of the National Academy of Sciences in Washington, D.C., in April 1949, and at the meeting of the American Society of Biological Chemists in Detroit in April, 1949.

This research was carried out with the aid of a grant from the United States Public Health Service. The authors are grateful to Professor Ray D. Owen, of the Biology Division of California Institute of Technology, for his helpful suggestions. We are indebted to Dr. Edward R. Evans, of Pasadena, Dr. Travis Winsor, of Los Angeles, and Dr. G. E. Burch, of the Tulane University School of Medicine, New Orleans, for their aid in obtaining the blood used in these experiments.

1949

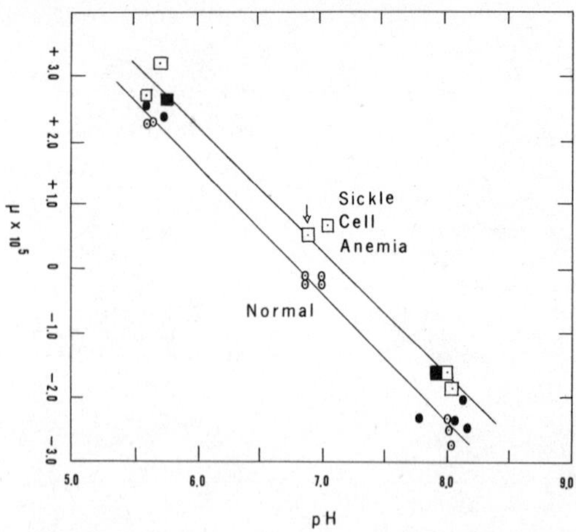

FIG. 1. Mobility (μ)-pH^{pH} curves for carbonmonoxyhemoglobins in phosphate buffers of 0.1 ionic strength. The black circles and black squares denote the data for experiments performed with buffers containing dithionite ion. The open square designated by the arrow represents an average value of 10 experiments on the hemoglobin of different individuals with sickle cell anemia. The mobilities recorded in this graph are averages of the mobilities in the ascending and descending limbs.

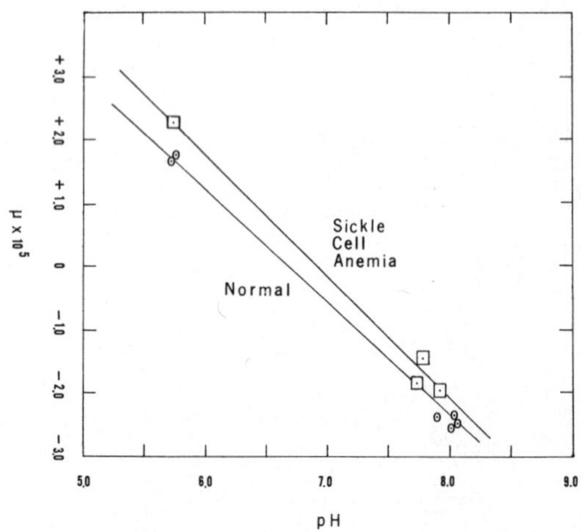

FIG. 2. Mobility (μ)-pH^{pH} curves for ferrohemoglobins in phosphate buffers of 0.1 ionic strength containing dithionite ion. The mobilities recorded in the graph are averages of the mobilities in the ascending and descending limbs.

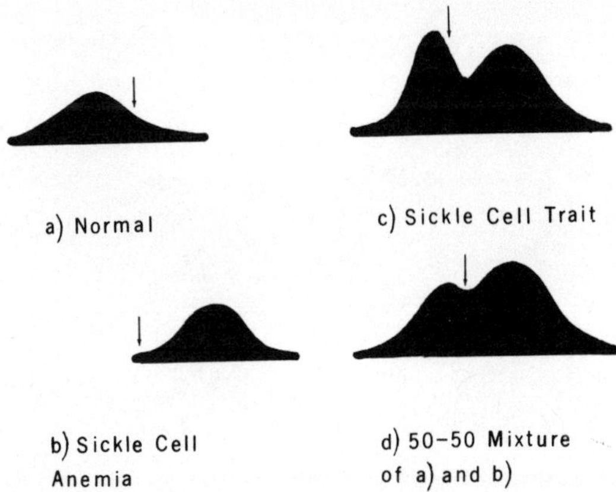

FIG. 3. Longsworth scanning diagrams of carbonmonoxyhemoglobins in phosphate buffer of 0.1 ionic strength and pH 6.90 taken after 20 hours' electrophoresis at a potential gradient of 4.73 volts/cm.

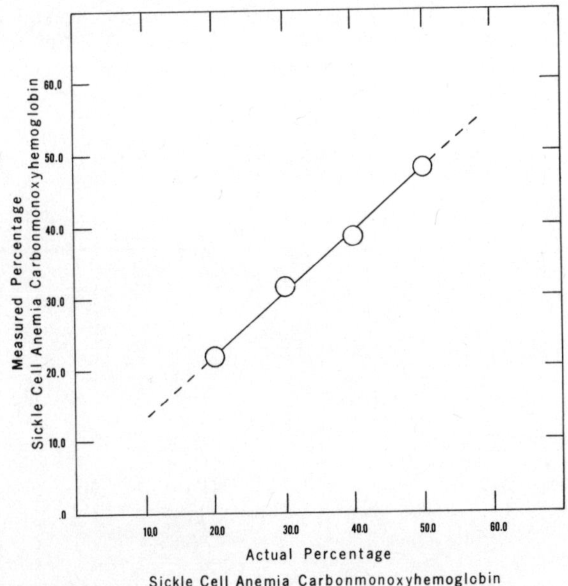

FIG. 4. The determination of the percent of sickle cell anemia carbonmonoxyhemoglobin in known mixtures of the protein with normal carbonmonoxyhemoglobin by means of electrophoretic analysis. The experiments were performed in a cacodylate–sodium chloride buffer described in the text.

REFERENCES

(1) Boyd, W. *Textbook of pathology.* (3rd ed.) Philadelphia: Lea and Febiger, 1938. P. 864.

(2) Diggs, L. W., Ahmann, C. F., and Bibb, J. *Ann. int. Med.*, 1933, 7, 769.

(3) Drabkin, D. L. *J. biol. Chem.*, 1946, 164, 703.

(4) Emmel, V. E. *Arch. int. Med.*, 1917, 20, 586.

(5) German, B. and Wyman, J., Jr. *J. biol. Chem.*, 1937, 117, 533.

(6) Hastings, A. B. *et al. J. biol. Chem.*, 1924, 60, 89.

(7) Longsworth, L. G. *Ann. N. Y. Acad. Sci.*, 1941, 41, 267.

(8) Neel, J. V. *Science,* 1949, 110, 64.

(9) Pauling, L., Pressman, D., and Campbell, D. H. *Physiol. Rev.*, 1943, 23, 203.

(10) Ponder, E. *Ann. N. Y. Acad. Sci.*, 1947, 48, 579.

(11) Sherman, I. J. *Bull. Johns Hopk. Hosp.*, 1940, 67, 309.

(12) Stern, K. G., Reiner, M. and Silber, R. H. *J. biol. Chem.*, 1945, 161, 731.

(13) Stern, C. *Science,* 1948, 108, 615.

(14) Swingle, S. M. *Rev. sci. Inst.*, 1947, 18, 128.

(15) Taliaferro, W. H. and Huck, J. G. *Genetics,* 1923, 8, 594.

(16) Tiselius, A. and Kabat, E. *J. exp. Med.*, 1939, 69, 119.

NOTES

[1] Occasionally small amounts (less than 5 percent of the total protein) of material with mobilities different from that of either kind of hemoglobin were observed in these uncrystallized hemoglobin preparations. According to the observations of Stern, Reiner, and Silber (12) a small amount of a component with a mobility smaller than that of oxyhemoglobin is present in human erythrocyte hemolyzates.

[2] The results obtained with carbonmonoxyhemoglobins with and without dithionite ion in the buffers indicate that the dithionite ion plays no significant role in the electrophoretic properties of the proteins. It is therefore of interest that ferrohemoglobin was found to have a lower isoelectric point in phosphate buffer than carbonmonoxyhemoglobin. Titration studies have indicated (5,6) that oxyhemoglobin (similar in electrophoretic properties to the carbonmonoxy compound) has a lower isoelectric point than ferrohemoglobin in the absence of other ions. These results might be reconciled by assuming that the ferrous iron of ferrohemoglobin forms complexes with phosphate ions which cannot be formed when the iron is combined with oxygen or carbon monoxide. We propose to continue the study of this phenomenon.

[3] The mobility data show that in 0.1 ionic strength cacodylate buffers the isoelectric points of the hemoglobins are increased about 0.5 pH unit over

their values in 0.1 ionic strength phosphate buffers. This effect is similar to that observed by Longsworth in his study of ovalbumin (7).

[4] The patterns were very slightly different in that the known mixture contained 1 percent more of the sickle cell anemia component than did the sickle cell trait material.

[5] We are indebted to Dr. M. Moskowitz, of the Chemistry Department, University of Berkeley, for performing the ultracentrifuge experiments for us.

V. M. Ingram

Gene Mutations in Human Haemoglobin: The Chemical Difference Between Normal and Sickle Cell Haemoglobin

The recognition, by Pauling, that sickle-cell anemia was a molecular disease, raised two plausible interpretations of the hemoglobin abnormality associated with the cell sickling. One view looked on abnormal hemoglobin as a change of its three-dimensional shape. The other view interpreted the abnormality as a substitution of one or more amino acids in the hemoglobin molecule. The hemoglobin molecule consists of four subunits. Two subunits, identical to one another, constitute the alpha chain. Two additional subunits, identical to one another but differing from the alpha protein, constitute the beta chain. The two alpha chains and the two beta chains form a clustered unit within which are located four heme groups. The heme groups contain an iron compound which picks up oxygen and releases it depending on the surrounding concentration of oxygen in the tissues. Ingram knew how important were Sanger's methods for working out the sequence of amino acids in a peptide. Hemoglobin, however, had many more amino acids than insulin and Sanger worked many years to achieve his structural analysis of that molecule. Fortunately, Ingram thought of using the enzyme trypsin to digest the hemoglobin molecules. This enzyme cleaves proteins at specific places (wherever lysine or arginine are present in the chain). The "tryptic digests" can be separated by chromatography. The streaks or "fingerprints" of the hemoglobin tryptic digests then can be cut out of the paper chromatogram and Sanger's techniques can be applied to analyze the amino acid content and sequence in them. By matching the fingerprints of hemoglobin from a sickle-cell-anemia patient with that obtained from normal

Reprinted from *Nature 180*:326–328, July–Dec., 1957, by permission of the publisher and the author.

blood, Ingram could determine how many regions of these two hemoglobin molecules differed from one another. To his surprise, Ingram found only one amino acid had changed among nearly three hundred amino acids that comprise the alpha and beta chains. Ingram's work immediately suggested that the gene controlling the beta chain (in which the replacement had occurred) was correspondingly altered as a point mutation by a single alteration of the nucleotide sequence.

I reported recently[1] that the globins of normal and sickle cell anaemia human haemoblogins differed only in a small portion of their polypeptide chains. I have now found that out of nearly 300 amino-acids in the two proteins, only one is different; one of the glutamic acid residues of normal haemoglobin is replaced by a valine residue in sickle cell anaemia haemoglobin. The latter is an abnormal protein which is inherited in a strictly Mendelian manner; it is now possible to show, for the first time, the effect of a single gene mutation as a change in one amino-acid of the haemoglobin polypeptide chain for the manufacture of which that gene is responsible.

In previous experiments,[1] tryptic digests of the two proteins had been prepared; the resulting mixtures of small peptides were separated on a sheet of paper, using electrophoresis in one direction and partition chromatography in the other. These paper chromatograms derived from the two proteins, which I had called "finger-prints", showed all peptides to have identical electrophoretic and chromatographic properties, except for one spot peptide No. 4. This occupied different positions in the "finger-prints" of normal (Hb *A*) and sickle cell anaemia (Hb *S*) haemoglobins, indicating that the difference between the two proteins was located there. I have now determined the chemical constitution of these No. 4 peptides derived from both haemoglobin *A* and *S*.

The haemoglobin *A* No. 4 peptide was prepared by elution from the neutral fraction of many "finger-prints", followed by cooled paper electrophoresis in pyridine/acetic acid/water (pH 3.6[2]) on Whatman No. 3 MM paper at 30 V./cm. for 75 min. The peptide was obtained as a well-separated band and eluted. The haemoglobin *S* No. 4 peptide could be produced in a fairly pure state by eluting the slowest positively charged band from an extended one-dimensional paper electrophoresis of the peptide

mixture in the pH 6.4 buffer.[1] In both cases qualitative amino-acid analysis by paper chromatography showed the presence of histidine, valine, leucine, threonine, proline, glutamic acid and lysine. There was more glutamic acid in the haemoglobin *A* peptide, but more valine in the haemoglobin *S* peptide. In view of the known specificity of trypsin,[3] it was to be expected that these peptides, obtained by tryptic hydrolysis, had lysine at the C-terminal end. This agreed with all the results from the partial acid hydrolysis studies as reported below.

Partial hydrolysis in 12 N hydrochloric acid at 37°C. for two or three days, followed by "finger-printing",[1] gave the peptides indicated in Fig. 1, and also free amino-acids, which are omitted from the figure. The N-terminal amino-acids of most of these peptides were determined by the fluoro-2,4-dinitrobenzene method.[4] Together with the amino-acid compositions, these fragments indicated the sequences of the No. 4 peptides of haemoglobins *A* and *S* shown in Fig. 1. The only ambiguity was the amino-acid following threonine. Here the relevant products from the hydrochloric acid splitting—threonyl-prolyl-glutamyl-glutamyl-lysine and threonyl-prolyl-valyl-glutamyl-lysine—were subjected, on paper strips, to a stepwise Edman degradation for two cycles.[4,5] The results indicated the sequence threonyl-prolyl- in both cases. The charge distribution of the two No. 4 peptides shown in Fig. 1 was deduced from the electrophoretic behaviour of the two peptides, especially in relation to the behaviour of the smaller split peptides. The only difference found between the two No. 4 peptides is that the first glutamic acid residue of the haemoglobin *A* peptide is replaced by valine in the haemoglobin *S* peptide.

It is known from X-ray crystallographic[6] and from chemical[7] studies that the human haemoglobin molecule of molecular weight 66,700 is composed of two identical half-molecules, each approximately 33,000. It is believed that this substitution, which occurs in each of the two identical half-molecules, constitutes the only chemical difference between normal and sickle cell anaemia haemoglobins. Certainly the haem groups of the two proteins are the same.[8] The fact that in each half-molecule a glutamic acid is replaced by the neutral amino-acid valine agrees with previous findings that the whole haemoglobin *S* molecule has two to three carboxyl groups fewer than the normal protein.[9,10] All the other peptides of the tryptic digest occupy identical, and characteristic,

positions in the two "finger-prints". Qualitative amino-acid analyses of these peptides have now been carried out, but have failed to reveal any differences between them. It would seem probable, therefore, that they have identical structures, leaving the two No. 4 peptides as the only ones that differ.

About 30 percent of the haemoglobin molecule is not susceptible to attack by trypsin and does not appear on the "finger-print". To eliminate the possibility that an additional difference resides in these large haemoglobin *A* and *S* fragments, they were digested with chymotrypsin, which attacks them readily. Again two peptide mixtures were obtained, which were examined both by "finger-printing" and by careful paper chromatography of the neutral peptides. No differences between them could be detected.

We owe to Pauling and his collaborators[9] the realization that sickle cell anaemia is an example of an inherited "molecular disease" and that it is due to an alteration leading to a protein which is by all criteria still a haemoglobin. It is now clear that, per half-molecule of haemoglobin, this change consists in a replacement of only one of nearly 300 amino-acids, namely, glutamic acid, by another, valine—a very small change indeed.

Differences between closely related proteins, involving only a very small number of amino-acids, are known; the clearest examples are the differences between horse, whale, sheep, pig and cattle insulins, which show changes in only one sequence of three amino-acids.[11] However, since these are inter-species differences, the genetic mechanism underlying them is by no means clear and cytoplasmic inheritance has not yet been ruled out. The abnormal human haemoglobins, on the other hand, are a group of very closely related proteins within the same species. It is certain that the inheritance of these proteins is Mendelian in character and occurs through the chromosomal genes. Neel[12] has shown that a single mutation step of such a gene, the one responsible for making haemoglobin, produces the new abnormal sickle cell anaemia haemoglobin. Previous investigations on the normal and the sickle cell anaemia protein could not decide whether the difference between them is due to a difference in folding of identical polypeptide chains or to a difference in the amino-acid sequences of the two chains. While there may also be changes in folding, it has now been definitely established that the amino-acid sequences of the two proteins differ, and differ at only one point. Thus it can be seen that an alteration in a Mendelian gene causes an

alteration in the amino-acid sequence of the corresponding polypeptide chain. In the case of sickle cell anaemia haemoglobin, this is the smallest alteration possible—only one amino-acid is affected—reflecting, presumably, a change in a very small portion of the haemoglobin gene. It is not known, but it may well be that this involves a replacement of no more than a single base-pair in the chain of the deoxyribonucleic acid of the gene.

It is well known that mutations lead to very small chemical differences between, for example, flower pigments.[13] It seems likely that these mutations produce first a change in a protein, in this case probably an enzyme, which in turn causes the production of a changed flower pigment. These enzymes, which have not yet been investigated for differences, stand in closer relationship to the gene than do the flower pigments themselves. The protein haemoglobin is just as close. It has therefore been called the first gene product, and is probably the first protein to be made by the gene.

The divisibility of genes in a virus was shown previously in bacteriophage by Benzer[14] and Streisinger,[15] who studied the effects of many different mutations of a gene on the behaviour of the virus. Such sub-units in genes have also been shown in *Aspergillus*[16] and certainly what one would expect on the basis of the widely accepted hypothesis of gene action; the sequence of base-pairs along the chain of nucleic acid provides the information which determines the sequence of amino-acids in the polypeptide chain for which the particular gene, or length of nucleic acid, is responsible. A substitution in the nucleic acids leads to a substitution in the polypeptide.

The abnormally low solubility of reduced haemoglobin *S*, which causes the sickling of the erythrocytes in the anaemia, is presumably a function of the charge distribution on the surface of the molecule. The replacement of two charged glutamic acid residues for two uncharged valines is presumably enough to alter this distribution towards one favouring abnormally easy crystallization.

It is hoped that similar studies of other abnormal human haemoglobins[18] will provide further insight into the effects of gene mutations.

Full details of this work will be published shortly elsewhere. I am indebted to Drs. S. Brenner and G. Seaman, Cambridge, for supplying blood from patients with homozygous sickle cell

anaemia. It is a pleasure to acknowledge the constant interest and encouragement shown by Drs. M. F. Perutz and F. H. C. Crick. Some of the enzymatic digestion and "finger-prints" were done by Mr. J. A. Hunt; Miss Rita Prior rendered invaluable assistance.

1957

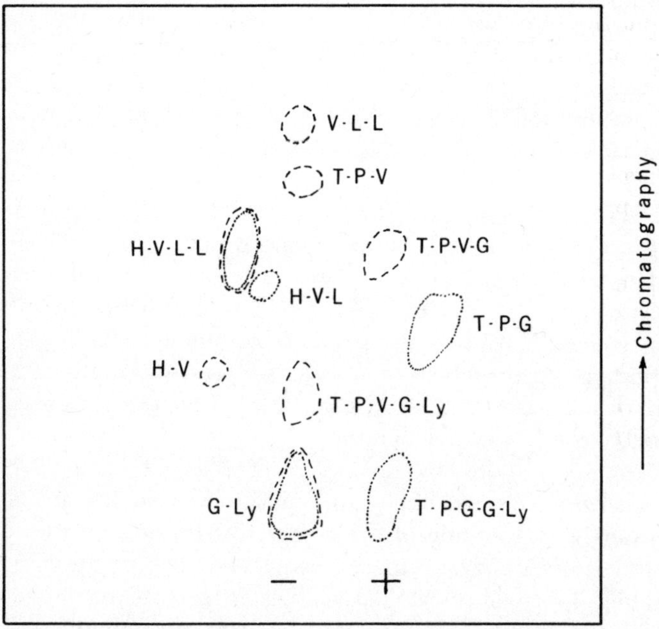

FIG. 1. Acid degradation and structure of the No. 4 peptides from haemoglobins *A* and *S*. Haemoglobin *A* (full lines): $\overset{++}{\text{His}}$-Val-Leu-Leu-Thr-Pro-$\overset{-}{\textit{Glu}}$-$\overset{-}{\text{Glu}}$-$\overset{+-}{\text{Lys}}$. Haemoglobin *S* (broken lines): $\overset{++}{\text{His}}$-Val-Leu-Leu-Thr-Pro-*Val*-$\overset{-}{\text{Glu}}$-$\overset{+-}{\text{Lys}}$.

REFERENCES

[1] Ingram, V. M. *Nature, 178,* 792 (1956).
[2] Michl, H. *Monatsh. Chem. 82,* 489 (1951).
[3] Sanger, F. and Tuppy, H., *Biochem. J. 49,* 463, 481 (1951). Sanger, F., and Thompson, E. O. P., *Biochem. J. 53,* 353, 356 (1953).
[4] Fraenkel-Conrat, H., Harris, J. I. and Levy, A. L., "Methods of Biochemical Analysis" *2,* 359 (1955).
[5] Edman, P., *Acta Chem. Scand., 4,* 277, 283 (1950).
[6] Perutz, M. F., Liquori, A. M., and Eirich, F., *Nature, 167,* 929 (1951).
[7] Schroeder, W. A., Rhinesmith, H. S., and Pauling, L. (in the press).
[8] Havinga, E. and Itano, H. A., *Proc. U.S. Nat. Acad. Sci., 39,* 65 (1953).
[9] Pauling, L., Itano, H. A., Singer, S. J. and Wells, I. C., *Science, 110,* 543 (1949).
[10] Scheinberg, I. H., Harris, R. S., and Spitzer, J. L., *Proc. U.S. Nat. Acad. Sci. 40,* 777 (1954).
[11] Harris, J. I., Sanger, F. and Naughton, M. A., *Arch. Biochem. Biophys., 65,* 427 (1956).
[12] Neel, J. V., *Science, 110,* 64 (1949).
[13] Haldane, J. B. S., "Biochemistry of Genetics" (Allen and Unwin, London, 1954).
[14] Benzer, S., in "The Chemical Basis of Heredity," edit. by McElroy, W. D., and Glass, B., 70 (Johns Hopkins Press, Baltimore, 1957).
[15] Streisinger, G., and Franklin, N. C., *Cold Spring Harbor Symp. Quant. Biol., 21,* 103 (1956).
[16] Pritchard, R. H., *Heredity, 9,* 343 (1955).
[17] Giles, N. H., Partridge, C. W. H., and Nelson, N. J., *Proc. U.S. Nat. Acad. Sci. 43,* 305 (1957).
[18] Itano, H. A., *Ann. Rev. Biochem., 25,* 331 (1956).

F. H. C. Crick

On Protein Synthesis

Of all the papers written on conceptual biology in the past twenty years, this essay by F. H. C. Crick ranks among the most influential and imaginative. Crick's thesis is that proteins do just about everything. Why then do we look to DNA as an important molecule? It is the *genetic code* which it must contain that fascinates Crick. In his analysis Crick suggests how protein synthesis can be accomplished through a genetic code. He also points out the philosophical assumption of *colinearity* inherent in genetic thought. The genetic map should correspond to the nucleotide map of DNA and the amino-acid map of proteins. Some of Crick's ideas, especially his genetic code, turned out to be wrong in detail. Nevertheless, the thinking, imagination, and clarity of Crick's thesis are well worth admiring.

I. INTRODUCTION

Protein synthesis is a large subject in a state of rapid development. To cover it completely in this article would be impossible. I have therefore deliberately limited myself here to presenting a broad general view of the problem, emphasizing in particular well-established facts which require explanation, and only selecting from recent work those experiments whose implications seem likely to be of lasting significance. Much very recent work, often of great interest, has been omitted because its implications are not clear. I have also tried to relate the problem to the other central problems of molecular biology—those of gene action and nucleic acid synthesis. In short, I have written for the biologist rather than the biochemist, the general reader rather than the specialist. More technical reviews have appeared recently by

Reprinted from *The Biological Replication of Macromolecules, Symposia of the Society for Experimental Biology XII*:138–163, 1958, by permission of the Society.

Borsook (1956), Spiegelman (1957), and Simkin and Work (1957b and this Symposium).

The importance of proteins

It is an essential feature of my argument that in biology proteins are uniquely important. They are not to be classed with polysaccharides, for example, which by comparison play a very minor role. Their nearest rivals are the nucleic acids. Watson said to me, a few years ago, "The most significant thing about the nucleic acids is that we don't know what they do." By contrast the most significant thing about proteins is that they can do almost anything. In animals proteins are used for structural purposes, but this is not their main role, and indeed in plants this job is usually done by polysaccharides. *The main function of proteins is to act as enzymes.* Almost all chemical reactions in living systems are catalysed by enzymes, and all known enzymes are proteins. It is at first sight paradoxical that it is probably easier for an organism to produce a new protein than to produce a new small molecule, since to produce a new small molecule one or more new proteins will be required in any case to catalyse the reactions.

I shall also argue that the main function of the genetic material is to control (not necessarily directly) the synthesis of proteins. There is a little direct evidence to support this, but to my mind the psychological drive behind this hypothesis is at the moment independent of such evidence. Once the central and unique role of proteins is admitted there seems little point in genes doing anything else. Although proteins can *act* in so many different ways, the way in which they are *synthesized* is probably uniform and rather simple, and this fits in with the modern view that gene action, being based upon the nucleic acids, is also likely to be uniform and rather simple.

Biologists should not deceive themselves with the thought that some new class of biological molecules, of comparable importance to the proteins, remains to be discovered. This seems highly unlikely. In the protein molecule Nature has devised a unique instrument in which an underlying simplicity is used to express great subtlety and versatility; it is impossible to see molecular biology in proper perspective until this peculiar combination of virtues has been clearly grasped.

II. THE PROBLEM

Elementary facts about proteins

(1) *Composition.* Simple (unconjugated) proteins break down on hydrolysis to amino acids. There is good evidence that in a native protein the amino acids are condensed into long polypeptide chains. A typical protein, of molecular weight about 25,000, will contain some 230 residues joined end-to-end to form a single polypeptide chain.

Two points are important. First, the actual chemical step required to form the covalent bonds of the protein is always the same, irrespective of the amino acid concerned, namely the formation of the peptide link with the elimination of water. Apart from minor exceptions (such as S—S links and, sometimes, the attachment of a prosthetic group) *all* the covalent links within a protein are formed in this way. Covalently, therefore, a protein is to a large extent a linear molecule (in the topological sense) and there is little evidence that the backbone is ever branched. From this point of view the cross-linking by S—S bridges is looked upon as a secondary process.

The second important point—and I am surprised that it is not remarked more often—is that only about twenty different *kinds* of amino acids occur in proteins, and that these same twenty occur, broadly speaking, in *all* proteins, of whatever origin—animal, plant or micro-organism. Of course not every protein contains every amino acid—the amino acid tryptophan, which is one of the rarer ones, does not occur in insulin, for example—but the majority of proteins contain at least one of each of the twenty amino acids. In addition all these twenty amino acids (apart from glycine) have the L configuration when they occur in genuine proteins.

There are a few proteins which contain amino acids not found elsewhere—the hydroxyproline of collagen is a good example—but in all such cases it is possible to argue that their presence is due to a modification of the protein after it has been synthesized or to some other abnormality. In Table I, I have listed the standard twenty amino acids believed to be of universal occurrence and also, in the last column, some of the exceptional ones. The assignment given in Table I might not be agreed by everyone, as

the evidence is incomplete, but more agreement could be found for this version than for any other. Curiously enough this point is slurred over by almost all biochemical textbooks, the authors of which give the impression that they are trying to include as many amino acids in their lists as they can, without bothering to distinguish between the magic twenty and the others. (But see a recent detailed review by Synge, 1957.)

TABLE I

The magic twenty amino acids found universally in proteins		Other amino acids found in proteins
Glycine	Asparagine	Hydroxproline[x]
Alanine	Glutamine	Hydroxlysine
Valine	Aspartic acid	Phosphoserine
Leucine	Glutamic acid	Diaminopimelic acid
Isoleucine	Arginine	Thyroxine and related molecules
Proline[x]	Lysine	
Phenylalanine	Histidine	
Tyrosine	Tryptophan	Cystine[xx]
Serine	Cysteine[xx]	
Threonine	Methionine	

[x] These are, of course, imino acids. This distinction is not made in the text.
[xx] This classification implies that all the cystine found in proteins is formed by the joining together of two cysteine molecules.

(2) *Homogeneity*. Not only is the composition of a given protein fixed, but we have every reason to believe that the exact order of the amino acid residues along the polypeptide chains is also rigidly determined: that each molecule of haemoglobin in your blood, for example, has exactly the same sequence of amino acids as every other one. This is clearly an overstatement; the mechanisms must make mistakes sometimes, and, as we shall see, there are also interesting exceptions which are under genetic control. Moreover, it is quite easy, in extracting a protein, to modify some of the molecules slightly without affecting the others, so that the "pure" protein may appear heterogeneous. The exact amount of "microheterogeneity" of proteins is controversial (see the review by Steinberg and Mihalyi, 1957), but

this should not blind one to the astonishing degree of homo geneity of most proteins.*

(3) *Structure.* In a native globular protein the polypeptide chain is not fully extended but is thrown into folds and super-folds, maintained by weak physical bonds, and in some cases by covalent—S—S—links and possibly some others. This folding is also thought to be at least broadly the same for each copy of a particular protein, since many proteins can be crystallized, though the evidence for *perfect* homogeneity of folding is perhaps rather weak.** As is well known, if this folding is destroyed by heat or other methods the protein is said to be "denatured". The biological properties of most proteins, especially the catalytic action of enzymes, must depend on the exact spatial arrangement of certain side-groups on the surface of the protein, and altering this arrangement by unfolding the polypeptide chains will destroy the biological specificity of the proteins.

(4) *Amino acid requirements.* If one of the twenty amino acids is supplied to a cell it can be incorporated into proteins; amino acids are certainly protein precursors. The only exceptions are amino acids like hydroxyproline, which are not among the magic twenty. The utilization of peptides is controversial but the balance of evidence is against the occurrence of peptide intermediates. (See the discussion by Simkin and Work, this Symposium.)

If, for some reason, one of the twenty amino acids is not available to the organism, protein synthesis stops. Moreover, the continued synthesis of those parts of the protein molecules which do not contain that amino acid appears not to take place. This can be demonstrated particularly clearly in bacteria, but it is also true of higher animals. If a meal is provided that lacks an essential amino acid it is no use trying to make up for this deficiency by providing it a few hours later.

Very little is known about the accuracy with which the amino acids are selected. One would certainly expect, for example, that the mechanism would occasionally put a valine into an isoleucine site, but exactly how often this occurs is not known. The impression one gets from the rather meagre facts at present available is that mistakes occur rather infrequently.

* The -globulins and other antibody molecules are exceptions to these generalizations. They are probably heterogeneous in folding and possibly to some extent in composition.

** See previous footnote.

In recent years it has been possible to introduce amino acid analogues into proteins by supplying the analogue under circumstances in which the amino acid itself is not easily available (see the review by Kamin and Handler, 1957). For example in *Escherichia coli* fluorophenylalanine has been incorporated in place of phenylalanine and tyrosine (Munier and Cohen, 1956) and it has even proved possible to replace completely the sulphur-containing amino acid methionine by its selenium analogue (Cohen and Cowie, 1957). Of the enzymes produced by the cell in these various ways some were active and some were inactive, as might have been expected.

(5) *Contrast with polysaccharides.* It is useful at this point to contrast proteins with polysaccharides to underline the differences between them. (I do not include nucleic acids among the polysaccharides.) Polysaccharides, too, are polymers, but each one is constructed from one, or at the most only about half-a-dozen kinds of monomer. Nevertheless many different monomers are found throughout Nature, some occurring here, some there. There is no standard set of monomers which is always used, as there is for proteins. Then polysaccharides are polydisperse—at least so far no monodisperse one has been found—and the order of their monomers is unlikely to be rigidly controlled, except in some very simple manner. Finally in those cases which have been carefully studied, such as starch, glycogen and hyaluronic acid, it has been found that the polymerization is carried out in a straightforward way by enzymes.

(6) *The genetics and taxonomy of proteins.* It is instructive to compare your own haemoglobin with that of a horse. Both molecules are indistinguishable in size. Both have similar amino acid compositions; similar but not identical. They differ a little electrophoretically, form different crystals, and have slightly different ends to their polypetide chains. All these facts are compatible with their polypeptide chains having similar amino acid sequences, but with just a few changes here and there.

This "family likeness" between the "same" protein molecules *from different species* is the rule rather than the exception. It has been found in almost every case in which it has been looked for. One of the best-studied examples is that of insulin, by Sanger and his co-workers (Brown, Sanger and Kitai, 1955; Harris, Sanger and Naughton, 1956), who have worked out the complete amino acid sequences for five different species, only two of which (pig

and whale) are the same. Interestingly enough the differences are all located in one small segment of one of the two chains.

Biologists should realize that before long we shall have a subject which might be called "protein taxonomy"—the study of the amino acid sequences of the proteins of an organism and the comparison of them between species. It can be argued that these sequences are the most delicate expression possible of the phenotype of an organism and that vast amounts of evolutionary information may be hidden away within them.

There is, however, nothing in the evidence presented so far to prove that these differences between species are under the control of Mendelian genes. It could be argued that they were transmitted cytoplasmically through the egg. On the other hand, there is much evidence that genes do affect enzymes, especially from work on micro-organisms such as *Neurospora* (see Wagner and Mitchell, 1955). The famous "one gene—one enzyme" hypothesis (Beadle, 1945) expresses this fact, although its truth is controversial (personally I believe it to be largely correct). However, in none of these cases has the protein (the enzyme, that is) ever been obtained pure.

There are a few cases where a Mendelian gene has been shown unambiguously to alter a protein, the most famous being that of human sickle-cell-anaemia haemoglobin, which differs electrophoretically from normal adult haemoglobin, as was discovered by Pauling and his co-workers (1949). Until recently it could have been argued that this was perhaps not due to a change in amino acid sequence, but only to a change in the folding. That the gene does in fact alter the amino acid sequence has now been conclusively shown by my colleague, Dr. Vernon Ingram. The difference is due to a valine residue occurring in the place of a glutamic acid one, and Ingram has suggestive evidence that this is the *only* change (Ingram, 1956, 1957). It may surprise the reader that the alteration of one amino acid out of a total of about 300 can produce a molecule which (when homozygous) is usually lethal before adult life but, for my part, Ingram's result is just what I expected.

The nature of protein synthesis

The basic dilemma of protein synthesis has been realized by many people, but it has been particularly aptly expressed by Dr. A. L. Dounce (1956):

> My interest in templates, and the conviction of their necessity, originated from a question asked me on my Ph.D. oral examination by Professor J. B. Sumner. He enquired how I thought proteins might be synthesized. I gave what seemed the obvious answer, namely, that enzymes must be responsible. Professor Sumner then asked me the chemical nature of enzymes, and when I answered that enzymes were proteins or contained proteins as essential components, he asked whether these enzyme proteins were synthesized by other enzymes and so on *ad infinitum*.
>
> The dilemma remained in my mind, causing me to look for possible solutions that would be acceptable, at least from the standpoint of logic. The dilemma, of course, involves the specificity of the protein molecule, which doubltess depends to a considerable degree on the sequence of amino acids in the peptide chains of the protein. The problem is to find a reasonably simple mechanism that could account for specific sequences without demanding the presence of an ever-increasing number of new specific enzymes for the synthesis of each new protein molecule.

It is thus clear that the synthesis of proteins must be radically different from the synthesis of polysaccharides, lipids, co-enzymes and other small molecules; that it must be relatively simple, and to a considerable extent uniform throughout Nature; that it must be highly specific, making few mistakes; and that in all probability it must be controlled at not too many removes by the genetic material of the organism.

The essence of the problem

A systematic discussion of our present knowledge of protein synthesis could usefully be set out under three headings, each dealing with a flux: the flow of energy, the flow of matter, and the flow of information. I shall not discuss the first of these here. I shall have something to say about the second, but I shall particularly emphasize the third—the flow of information.

By information I mean the specification of the amino acid sequence of the protein. It is conventional at the moment to consider separately the synthesis of the polypeptide chain and its folding. It is of course possible that there is a special mechanism for folding up the chain, but the more likely hypothesis is that the *folding is simply a function of the order of the amino acids,* provided it takes place as the newly formed chain comes off the template. I think myself that this latter idea may well be correct,

though I would not be surprised if exceptions existed, especially the -globulins and the adaptive enzymes.

Our basic handicap at the moment is that we have no easy and precise technique with which to study how proteins are folded, whereas we can at least make some experimental approach to amino acid sequences. For this reason, if for no other, I shall ignore folding in what follows and concentrate on the determination of sequences. It is as well to realize, however, that the idea that the two processes can be considered separately is in itself an assumption.

The actual chemical step by which any two amino acids (or activated amino acids) are joined together is probably always the same, and may well not differ significantly from any other biological condensation. The unique feature of protein synthesis is that only a single standard set of twenty amino acids can be incorporated, and that for any particular protein the *amino acids must be joined up in the right order.* It is this problem, the problem of "sequentialization", which is the crux of the matter, though it is obviously important to discover the exact chemical steps which lead up to and permit the crucial act of sequentialization.

As in even a small bacterial cell there are probably a thousand different kinds of protein, each containing some hundreds of amino acids in its own rigidly determined sequence, the amount of hereditary information required for sequentialization is quite considerable.

III. RECENT EXPERIMENTAL WORK

The role of the nucleic acids

It is widely believed (though not by everyone) that the nucleic acids are in some way responsible for the control of protein synthesis, either directly or indirectly. The actual evidence for this is rather meagre. In the case of deoxyribonucleic acid (DNA) it rests partly on the T-even bacteriophages, since it has been shown, mainly by Hershey and his colleagues, that whereas the DNA of the infecting phage penetrates into the bacterial cell almost all the protein remains outside (see the review by Hershey, 1956); and also on Transforming Factor, which appears to be pure DNA, and which in at least one case, that of the enzyme mannitol phosphate dehydrogenase, controls the synthesis of a

protein (Marmur and Hotchkiss, 1955). There is also the indirect evidence that DNA is the most constant part of the genetic material, and that genes control proteins. Finally there is the very recent evidence, mainly due to the work of Benzer on the rII locus of bacteriophage, that the functional gene—the "cistron" of Benzer's terminology—consists of many sites arranged strictly *in a linear order* (Benzer, 1957) as one might expect if a gene controls the order of the amino acids in some particular protein.

As is well known, the correlation between ribonucleic acid (RNA) and protein synthesis was originally pointed out by Brachet and by Caspersson. Is there any more direct evidence for this connexion? In particular is there anything to support the idea that the sequentialization of the amino acids is controlled by the RNA?

The most telling evidence is the recent work on tobacco mosaic virus. A number of strains of the virus are known, and it is not difficult to show (since the protein sub-unit of the virus is small) that they differ in amino acid composition. Some strains, for example, have histidine in their protein, whereas others have none. Two very significant experiments have been carried out. In one, as first shown by Gierer and Schramm (1956), the RNA of the virus alone, although completely free of protein, appears to be infective, though the infectivity is low. In the other, first done by Fraenkel-Conrat, it has proved possible to separate the RNA from the protein of the virus and then recombine them to produce virus again. In this case the infectivity is comparatively high, though some of it is usually lost. If a recombined virus is made using the RNA of one strain and the protein of another, and then used to infect the plant, the new virus produced in the plant resembles very closely *the strain from which the RNA was taken.* If this strain had a protein which contained no histidine then the offspring will have no histidine either, although the plant had never been in contact with this particular protein before but only with the RNA from that strain. In other words *the viral RNA appears to carry at least part of the information which determines the composition of the viral protein.* Moreover the viral protein which was used to infect the cell was not copied to any appreciable extent (Fraenkel-Conrat, 1956).

It has so far not proved possible to carry out this experiment—a model of its kind—in any other system, although very recently

it has been claimed that for two animal viruses the RNA alone appears to be infective.

Turnover experiments have shown that while the labelling of DNA is homogeneous that of RNA is not. The RNA of the cell is partly in the nucleus, partly in particles in the cytoplasm and partly as the "soluble" RNA of the cell sap; many workers have shown that all these three fractions turn over differently. It is very important to realize in any discussion of the role of RNA in the cell that it is very inhomogeneous metabolically, and probably of more than one type.

The site of protein synthesis

There is no known case in Nature in which protein synthesis proper (as opposed to protein modification) occurs outside cells, though, as we shall see later, a certain amount of protein can probably be synthesized using broken cells and cell fragments. The first question to ask, therefore, is whether protein synthesis can take place in the nucleus, in the cytoplasm, or in both.

It is almost certain that protein synthesis can take place in the cytoplasm without the presence of the nucleus, and it is probable that it can take place to some extent in the nucleus by itself (see the review by Brachet and Chantrenne, 1956). Mirsky and his colleagues (see the review by Mirsky, Osawa and Allfrey, 1956) have produced evidence that some protein synthesis can occur in isolated nuclei, but the subject is technically difficult and in this review I shall quite arbitrarily restrict myself to protein synthesis in the cytoplasm.

In recent years our knowledge of the structure of the cytoplasm has enormously increased, due mainly to the technique of cutting thin sections for the electron microscope. The cytoplasm of many cells contains an "endoplasmic reticulum" of double membranes, consisting mainly of protein and lipid (see the review of Palade, 1956). On one side of each membrane appear small electron-dense particles (Palade, 1955). Biochemical studies (Palade and Siekevitz, 1956; among others) have shown that these particles, which are about 100–200 Å. in diameter, consist almost entirely of protein and RNA, in about equal quantities. Moreover the major part of the RNA of the cell is found in these particles.

When such a cell is broken open and the contents fractionated by centrifugation, the particles, together with fragments of the

endoplasmic reticulum, are found in the "microsome" fraction, and for this reason I shall refer to them as microsomal particles.

These microsomal particles are found in almost all cells. They are particularly common in cells which are actively synthesizing protein whereas the endoplasmic reticulum is most conspicuously present in (mammalian) cells which are secreting very actively. Thus both the cells of the pancreas and those of an ascites tumour contain large quantities of microsomal particles, but the tumour has little endoplasmic reticulum, whereas the pancreas has a lot. Moreover, there is no endoplasmic reticulum in bacteria.

On the other hand particles of this general description have been found in plant cells (Ts'o, Bonner and Vinograd, 1956), in yeast, and in various bacteria (Schachman, Pardee and Stanier, 1953); in fact in all cells which have been examined for them.

These particles have been isolated from various cells and examined in the ultra-centrifuge (Petermann, Mizen and Hamilton, 1952; Schachman *et al.*, 1953; among others). The remarkable fact has emerged that they do not have a continuous distribution of sedimentation constants, but usually fall into several well-defined groups. Moreover some of the particles are probably simple aggregates of the others (Petermann and Hamilton, 1957). This uniformity suggests immediately that the particles, which have "molecular weights" of a few million, have a definite structure. They are, in fact, reminiscent of the small spherical RNA-containing viruses, and Watson and I have suggested that they may have a similar type of substructure (Crick and Watson, 1956).

Biologists should contrast the older concept of *microsomes* with the more recent and significant one of *microsomal particles*. Microsomes came in all sizes, and were irregular in composition; microsomal particles occur in a few sizes only, have a more fixed composition and a much higher proportion of RNA. It was hard to identify microsomes in all cells, whereas RNA-rich particles appear to occur in almost every kind of cell. In short, microsomes were rather a mess, whereas microsomal particles appeal immediately to one's imagination. It will be surprising if they do not prove to be of fundamental importance.

It should be noted, however, that Simpson and his colleagues (Simpson and McLean, 1955; Simpson, McLean, Cohn and Brandt, 1957) have reported that protein synthesis can take place in mitochondria. It is known that mitochondria contain RNA, and it would be of great interest to know whether this RNA is

in some kind of particle. Mitochondria are, of course, very widely distributed but they do not occur in lower forms such as bacteria. Similar remarks about RNA apply to the reported incorporation in chloroplasts (Stephenson, Thimann and Zamecnik, 1956).

Microsomal particles and protein synthesis

It has been shown by the use of radioactive amino acids that during protein synthesis the amino acids appear to flow through the microsomal particles. The most striking experiments are those of Zamecnik and his co-workers on the livers of growing rats (see review by Zamecnik *et al.,* 1956).

Two variations of the experiment were made. In the first the rat was given a rather large intravenous dose of a radioactive amino acid. After a predetermined time the animal was sacrificed, the liver extracted, its cells homogenized and the contents fractionated. It was found that the microsomal particle fraction was very rapidly labelled to a constant level.

In the second a very small shot of the radioactive amino acid was given, so that the liver received only a pulse of labelled amino acid, since this small amount was quickly used up. In this case the radioactivity of the microsomal particles rose very quickly *and then fell away*. Making plausible assumptions Zamecnik and his colleagues have shown that this behaviour is what one would expect if most of the protein of the microsomal particles were metabolically inert, but 1 or 2% was turning over very rapidly, say within a minute or so.

Very similar results have been obtained by Rabinovitz and Olson (1956, 1957) using intact mammalian cells, in this case rabbit reticulocytes. They have also been able to show that the label passed into a well-defined globular protein, namely haemoglobin. Experiments along the same general lines have also been reported for liver by Simkin and Work (1957a).

We thus have direct experimental evidence that the microsomal particles are associated with protein synthesis, though the precise role they play is not clear.

Activating enzymes

It now seems very likely that the first step in protein synthesis is the activation of each amino acid by means of its special

"activating enzyme". The activation requires ATP, and the evidence suggests that the reaction is

amino acid + ATP = AMP – amino acid + pyrophosphate

The activated amino acid, which is probably a mixed anhydride of the form

```
                   O
                   |
     NH2     O—P—O—Ribose—Adenine
      |     /      |
  R—C—C          O
      |     \\
      H       O
```

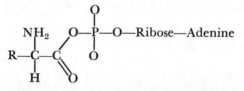

in which the carboxyl group of the amino acid is phosphorylated, appears to be tightly bound to its enzyme and is not found free in solution.

These enzymes were first discovered in the cell-sap fraction of rat liver cells by Hoagland (Hoagland, 1955; Hoagland, Keller and Zamecnik, 1956) and in yeast by Berg (1956). They have been shown by DeMoss and Novelli (1956) to be widely distributed in bacteria, and it is surmised that they occur in all cells engaged in protein synthesis. Recently Cole, Coote and Work (1957) have reported their presence in a variety of tissues from a number of animals.

So far good evidence has been found for this reaction for about half the standard twenty amino acids, but it is believed that further research will reveal the full set. Meanwhile Davie, Koningsberger and Lipmann (1956) have purified the tryptophan-activating enzyme. It is specific for trytophan (and certain tryptophan analogues) and will only handle the L-isomer. Isolation of the tyrosine enzyme has also been briefly reported (Koningsberger, van de Ven and Overbeck, 1957; Schweet, 1957).

The properties of these enzymes are obviously of the greatest interest, and much work along these lines may be expected in the near future. For example, it has been shown that the tryptophan-activating enzyme contains what is probably a derivative of guanine (perhaps GMP) very tightly bound. It is possible to remove it, however, and to show that its presence is not necessary for the primary activation step. Since the enzyme is probably involved in the next step in protein synthesis it is naturally suspected that the guanine derivative is also required for this reaction, whatever it may be.

In vitro *incorporation*

In order to study the relationship between the activating enzymes and the microsomal particles it has proved necessary to break open the cells and work with certain partly purified fractions. Unfortunately it is rare to obtain substantial net protein synthesis from such systems, and there is a very real danger that the incorporation of the radioactivity does not represent true synthesis but is some kind of partial synthesis or exchange reaction. This distinction has been clearly brought out by Gale (1953). The work to be described, therefore, has to be accepted with reservations. (See the remarks of Simkin and Work, this Symposium.) It has been shown, however, in the work described below, that the amino acid is incorporated into true peptide linkage.

Again the significant results were first obtained by Zamecnik and his co-workers (reviewed in Zamecnik *et al.,* 1956). The requirements so far known appear to fall into two parts:

(1) The activation of the amino acids for which, in addition to the labelled amino acid, one requires the "pH 5" fraction, containing the activating enzymes, ATP and (usually) an ATP-generating system. There appears to be no requirement for any of the pyrimidine or guanine nucleotides.

(2) The transfer to the microsomal particles. For this one requires the previous system plus GTP or GDP (Keller and Zamecnik, 1956) and of course the microsomal particles; the endoplasmic reticulum does not appear to be necessary (Littlefield and Keller, 1957).

Hultin and Beskow (1956) have reported an experiment which shows clearly that the amino acids become bound in some way. They first incubate the mixture described in (1) above. They then add a great excess of *un*labelled amino acid before adding the microsomal particles. Nevertheless some of the labelled amino acid is incorporated into protein, showing that it was in some place where it could not readily be diluted.

Very recently an intermediate reaction has been suggested by the work of Hoagland, Zamecnik and Stephenson (1957), who have discovered that in the first step the "soluble" RNA contained in the "pH 5" fraction became labelled with the radioactive amino acid. The bond between the amino acid and the RNA appears to be a covalent one. This labelled RNA can be extracted, purified,

and then added to the microsomal fraction. In the presence of GTP the labelled amino acid is transferred from the soluble RNA to microsomal protein. This very exciting lead is being actively pursued.

Many other experiments have been carried out on cell-free systems, in particular by Gale and Folkes (1955) and by Spiegelman (see his review, 1957), but I shall not describe them here as their interpretation is difficult. It should be mentioned that Gale (reviewed in Gale, 1956) has isolated from hydrolysates of commercial-yeast RNA a series of fractions which greatly increase amino acid incorporation. One of them, the so-called "glycine incorporation factor" has been purified considerably, and an attempt is being made to discover its structure.

RNA turnover and protein synthesis

From many points of view it seems highly likely that the *presence* of RNA is essential for cytoplasmic protein synthesis, or at least for specific protein synthesis. It is by no means clear, however, that the *turnover* of RNA is required.

In discussing this a strong distinction must be made between cells which are growing, and therefore producing new microsomal particles, and cells which are synthesizing without growth, and in which few new microsomal particles are being produced.

This is a difficult aspect of the subject as the evidence is to some extent conflicting. It appears reasonably certain that not *all* the RNA in the cytoplasm is turning over very rapidly—this has been shown, for example, by the Hokins (1954) working on amylase synthesis in slices of pigeon pancreas, though in the light of the recent work of Straub (this Symposium) the choice of amylase was unfortunate. On the other hand, Pardee (1954) has demonstrated that mutants of *Escherichia coli* which require uracil or adenine cannot synthesize β-galactosidase unless the missing base is provided.

Can RNA be synthesized without protein being synthesized? This can be brought about by the use of *chloramphenicol*. In bacterial systems chloramphenicol stops protein synthesis dead, but allows "RNA" synthesis to continue. A very interesting phenomenon has been uncovered in *E. coli* by Pardee and Prestidge (1956), and by Gros and Gros (1956). If a mutant is used which requires, say, leucine, then when the external supply of leucine is

exhausted both protein and RNA synthesis cease. If now chloramphenicol is added there is no effect, but if in addition the cells are given a small amount of leucine then rapid RNA synthesis takes place. If the chloramphenicol is removed, so that protein synthesis restarts, then this leucine is built into proteins and then, once again, the synthesis of both protein and RNA is prevented. In other words it appears as if "free" leucine (i.e., not bound into proteins) is required for RNA synthesis. This effect is not peculiar to leucine and has already been found for several amino acids and in several different organisms (Yčas and Brawerman, 1957).

As a number of people have pointed out, the most likely interpretation of these results is that protein and RNA require *common intermediates* for their synthesis, consisting in part of amino acids and in part of RNA components such as nucleotides. This is a most valuable idea; it explains a number of otherwise puzzling facts and there is some hope of getting close to it experimentally.

For completeness it should be stated that Anfinsen and his co-workers have some evidence that proteins are not produced from (activated) amino acids in a single step (see the review by Steinberg, Vaughan and Anfinsen, 1956), since they find unequal labelling between the same amino acid at different points on the polypeptide chain, but this interpretation of their results is not accepted by all workers in the field. This is discussed more fully by Simkin and Work (this Symposium).

Summary of experimental work

Both DNA and RNA have been shown to carry some of the specificity for protein synthesis. The RNA of almost all types of cell is found mainly in rather uniform, spherical, virus-like particles in the cytoplasm, known as microsomal particles. Most of their protein and RNA is metabolically rather inert. Amino acids, on their way into protein, have been shown to pass rapidly through these particles.

An enzyme has been isolated which, when supplied with tryptophan and ATP, appears to form an activated tryptophan. There is evidence that there exist similar enzymes for most of the other amino acids. These enzymes are widely distributed in Nature.

Work on cell fractions is difficult to interpret but suggests that the first step in protein synthesis involves these enzymes, and

that the subsequent transfer of the activated amino acids to the microsomal particles requires GTP. The soluble RNA also appears to be involved in this process.

Whereas the presence of RNA is probably required for true protein synthesis its rapid turnover does not appear to be necessary, at least not for all the RNA. There is suggestive evidence that common intermediates, containing both amino acids and nucleotides, occur in protein synthesis.

IV. IDEAS ABOUT PROTEIN SYNTHESIS

It is an extremely difficult matter to present current ideas about protein synthesis in a stimulating form. Many of the general ideas on the subject have become rather stale, and an extended discussion of the more detailed theories is not suitable in a paper for non-specialists. I shall therefore restrict myself to an outline sketch of my own ideas on cytoplasmic protein synthesis, some of which have not been published before. Finally I shall deal briefly with the problem of "coding".

General principles

My own thinking (and that of many of my colleagues) is based on two general principles, which I shall call the Sequence Hypothesis and the Central Dogma. The direct evidence for both of them is negligible, but I have found them to be of great help in getting to grips with these very complex problems. I present them here in the hope that others can make similar use of them. Their speculative nature is emphasized by their names. It is an instructive exercise to attempt to build a useful theory without using them. One generally ends in the wilderness.

The Sequence Hypothesis

This has already been referred to a number of times. In its simplest form it assumes that the specificity of a piece of nucleic acid is expressed solely by the sequence of its bases, and that this sequence is a (simple) code for the amino acid sequence of a particular protein.

This hypothesis appears to be rather widely held. Its virtue is that it unites several remarkable pairs of generalizations: the central biochemical importance of proteins and the dominating

biological role of genes, and in particular of their nucleic acid; the linearity of protein molecules (considered covalently) and the genetic linearity within the functional gene, as shown by the work of Benzer (1957) and Pontecorvo (this Symposium); the simplicity of the composition of protein molecules and the simplicity of the nucleic acids. Work is actively proceeding in several laboratories, including our own, in an attempt to provide more direct evidence for this hypothesis.

The Central Dogma

This states that once "information" has passed into protein it *cannot get out again*. In more detail, the transfer of information from nucleic acid to nucleic acid, or from nucleic acid to protein may be possible, but transfer from protein to protein, or from protein to nucleic acid is impossible. Information means here the *precise* determination of sequence, either of bases in the nucleic acid or of amino acid residues in the protein.

This is by no means universally held—Sir Macfarlane Burnet, for example, does not subscribe to it—but many workers now think along these lines. As far as I know it has not been *explicitly* stated before.

Some ideas on cytoplasmic protein synthesis

From our assumptions it follows that there must be an RNA template in the cytoplasm. The obvious place to locate this is in the microsomal particles, because their uniformity of size suggests that they have a regular structure. It also follows that the synthesis of at least some of the microsomal RNA must be under the control of the DNA of the nucleus. This is because the amino acid sequence of the human haemoglobins, for example, is controlled at least in part by a Mendelian gene, and because spermatozoa contain no RNA. Therefore, granted our hypothesis, the information must be carried by DNA.

What can we guess about the structure of the microsomal particle? On our assumptions the protein component of the particles can have no significant role in determining the amino acid sequence of the proteins which the particles are producing. We therefore assume that their main function is a structural one, though the possibility of some enzyme activity is not excluded. The simplest model then becomes one in which each particle is

made of the same protein, or proteins, as every other one in the cell, and has the same basic *arrangement* of the RNA, but that different particles have, in general, different base-sequences in their RNA, and therefore produce different proteins. This is exactly the type of structure found in tobacco mosaic virus, where the interaction between RNA and protein does not depend upon the sequence of bases of the RNA (Hart and Smith, 1956). In addition Watson and I have suggested (Crick and Watson, 1956), by analogy with the spherical viruses, that the protein of microsomal particles is probably made of many identical sub-units arranged with cubic symmetry.

On this oversimplified picture, therefore, the microsomal particles in a cell are all the same (except for the base-sequence of their RNA) and are metabolically rather inert. The RNA forms the template and the protein supports and protects the RNA.

This idea is in sharp contrast to what one would naturally assume at first glance, namely that the protein of the microsomal particles consists entirely of protein being synthesized. The surmise that most of the protein is structural was derived from considerations about the structure of virus particles and about coding; it was independent of the direct experimental evidence of Zamecnik and his colleagues that only a small fraction of the protein turns over rapidly, so that this agreement between theory and experiment is significant, as far as it goes.

It is obviously of the first importance to know how the RNA of the particles is arranged. It is a natural deduction from the Sequence Hypothesis that the RNA backbone will follow as far as possible a spatially regular path, in this case a helix, essentially because the fundamental operation of making the peptide link is always the same, and we therefore expect any template to be spatially regular.

Although we do not yet know the structure of isolated RNA (which may be an artifact) we do know that a pair of RNA-like molecules can under some circumstances form a double-helical structure, somewhat similar to DNA, because Rich and Davies (1956) have shown that when the two polyribotides, polyadenylic acid and polyuridylic acid (which have the same backbone as RNA) are mixed together they wind round one another to form a double helix, presumably with their bases paired. It would not be surprising, therefore, if the RNA backbone took up a helical configuration similar to that found for DNA.

This suggestion is in contrast to the idea that the RNA and protein interact in a complicated, irregular way to form a "nucleoprotein". As far as I know there is at the moment no direct experimental evidence to decide between these two points of view.

However, even if it turns out that the RNA is (mainly) helical and that the structural protein is made of sub-units arranged with cubic symmetry it is not at all obvious how the two could fit together. In abstract terms the problem is how to arrange a long fibrous object inside a regular polyhedron. It is for this reason that the structure of the spherical viruses is of great interest in this context, since we suspect that the same situation occurs there; moreover they are at the moment more amenable to experimental attack. A possible arrangement, for example, is one in which the axes of the RNA helices run radially and clustered in groups of five, though it is always possible that the arrangement of the RNA is irregular.

It would at least be of some help if the approximate location of the RNA in the microsomal particles could be discovered. Is it on the outside or the inside of the particles, for example, or even both? Is the microsomal particle a rather open structure, like a sponge, and if it is what size of molecule can diffuse in and out of it? Some of these points are now ripe for a direct experimental attack.

The adaptor hypothesis

Granted that the RNA of the microsomal particles, regularly arranged, is the template, how does it direct the amino acids into the correct order? One's first naive idea is that the RNA will take up a configuration capable of forming twenty different "cavities", one for the side-chain of each of the twenty amino acids. If this were so one might expect to be able to play the problem backwards—that is, to find the configuration of RNA by trying to form such cavities. All attempts to do this have failed, and on physical-chemical grounds the idea does not seem in the least plausible (Crick, 1957a). Apart from the phosphate-sugar backbone, which we have assumed to be regular and perhaps linked to the structural protein of the particles, RNA presents mainly a sequence of sites where hydrogen bonding could occur. One would expect, therefore, that whatever went on to the template in a *specific* way did so by forming hydrogen bonds. It is therefore a natural

hypothesis that the amino acid is carried to the template by an "adaptor" molecule, and that the adaptor is the part which actually fits on to the RNA. In its simplest form one would require twenty adaptors, one for each amino acid.

What sort of molecules such adaptors might be is anybody's guess. They might, for example, be proteins, as suggested by Dounce (1952) and by the Hokins (1954) though personally I think that proteins, being rather large molecules, would take up too much space. They might be quite unsuspected molecules, such as amino sugars. But there is one possibility which seems inherently more likely than any other—that they might contain nucleotides. This would enable them to join on to the RNA template by the same "pairing" of bases as is found in DNA, or in polynucleotides.

If the adaptors were small molecules one would imagine that a separate enzyme would be required to join each adaptor to its own amino acid and that the specificity required to distinguish between, say, leucine, isoleucine and valine would be provided by these enzyme molecules instead of by cavities in the RNA. Enzymes, being made of protein, can probably make such distinctions more easily than can nucleic acid.

An outline picture of the early stages of protein synthesis might be as follows: the template would consist of perhaps a single chain of RNA. (As far as we know a single isolated RNA backbone has no regular configuration [Crick, 1957b] and one has to assume that the backbone is supported in a helix of the usual type by the structural protein of the microsomal particles.) Alternatively the template might consist of a pair of chains. Each adaptor molecule containing, say, a di- or trinucleotide would be joined to its own amino acid by a special enzyme. These molecules would then diffuse to the microsomal particles and attach to the proper place on the bases of the RNA by base-pairing, so that they would then be in a position for polymerization to take place.

It will be seen that we have arrived at the idea of common intermediates without using the direct experimental evidence in their favour; but there is one important qualification, namely that the nucleotide part of the intermediates must be specific for each amino acid, at least to some extent. It is not sufficient, from this point of view, merely to join adenylic acid to each of the twenty amino acids. Thus one is led to suppose that after the activating step, discovered by Hoagland and described earlier, some other

more specific step is needed before the amino acid can reach the template.

The soluble RNA

If trinucleotides, say, do in fact play the role suggested here their synthesis presents a puzzle, since one would not wish to invoke too many enzymes to do the job. It seems to me plausible, therefore, that the twenty different adaptors may be synthesized by the *breakdown* of RNA, probably the "soluble" RNA. Whether this is in fact the same action which the "activating enzymes" carry out (presumably using GTP in the process) remains to be seen.

From this point of view the RNA with amino acids attached reported recently by Hoagland, Zamecnik and Stephenson (1957), would be a halfway step in this process of breaking the RNA down to trinucleotides and joining on the amino acids. Of course alternative interpretations are possible. For example, one might surmise that numerous amino acids become attached to this RNA and then proceed to polymerize, perhaps inside the microsomal particles. I do not like these ideas, because the supernatant RNA appears to be too short to code for a complete polypeptide chain, and yet too long to join on to template RNA (in the microsomal particles) by base-pairing, since it would take too great a time for a piece of RNA twenty-five nucleotides long, say, to diffuse to the correct place in the correct particles. If it were only a trinucleotide on the other hand, there would be many different "correct" places for it to go to (wherever a valine was required, say), and there would be no undue delay.

Leaving theories on one side, it is obviously of the greatest interest to know what molecules actually pass from the "pH 5 enzymes" to the microsomal particles. Are they small molecules, free in solution, or are they bound to protein? Can they be isolated? This seems at the moment to be one of the most fruitful points at which to attack the problem.

Subsequent steps

What happens after the common intermediates have entered the microsomal particles is quite obscure. Two views are possible, which might be called the Parallel Path and the Alternative Path

theories. In the first an intermediate is used to produce both protein and RNA at about the same time. In the second it is used to produce either protein, or RNA, but not both. If we knew the exact nature of the intermediates we could probably decide which of the two was more likely. At the moment there seems little reason to prefer one theory to the other.

The details of the polymerization step are also quite unknown. One tentative theory, of the Parallel Path type, suggests that the intermediates first polymerize to give an RNA molecule with amino acids attached. This process removes it from the template and it diffuses outside the microsomal particle. There the RNA folds to a new configuration, and the amino acids become polymerized to form a polypeptide chain, which folds up as it is made to produce the finished protein. The RNA, now free of amino acids, is then broken down to produce fresh intermediates. A great variety of theories along these lines can be constructed. I shall not discuss these further here, nor shall I describe the various speculations about the actual details of the chemical steps involved.

Two types of RNA

It is an essential feature of these ideas that there should be *at least two types of RNA in the cytoplasm.* The first, which we may call "template RNA" is located inside the microsomal particles. It is probably synthesized in the nucleus (Goldstein and Plaut, 1955) under the direction of DNA, and carries the information for sequentialization. It is metabolically inert during protein synthesis, though naturally it may show turnover whenever microsomal particles are being synthesized (as in growing cells), or breaking down (as in certain starved cells).

The other postulated type of RNA, which we may call "metabolic RNA", is probably synthesized (from common intermediates) in the microsomal particles, where its sequence is determined by base-pairing with the template RNA. Once outside the microsomal particles it becomes "soluble RNA" and is constantly being broken down to form the common intermediates with the amino acids. It is also possible that some of the soluble RNA may be synthesized in a random manner in the cytoplasm; perhaps in bacteria, by the enzyme system of Grunberg-Manago and Ochoa (1955).

One might expect that there would also be metabolic RNA in the nucleus. The existence of these different kinds of RNA may well explain the rather conflicting data on RNA turnover.

The coding problem

So much for biochemical ideas. Can anything about protein synthesis be discovered by more abstract arguments? If, as we have assumed, the sequence of bases along the nucleic acid determines the sequence of amino acids of the protein being synthesized, it is not unreasonable to suppose that this interrelationship is a simple one, and to invent abstract descriptions of it. This problem of how, in outline, the sequence of four bases "codes" the sequence of the twenty amino acids is known as the coding problem. It is regarded as being independent of the biochemical steps involved, and deals only with the transfer of information.

This aspect of protein synthesis appeals mainly to those with a background in the more sophisticated sciences. Most biochemists, in spite of being rather fascinated by the problem, dislike arguments of this kind. It seems to them unfair to construct theories without adequate experimental facts. Cosmologists, on the other hand, appear to lack such inhibitions.

The first scheme of this kind was put forward by Gamow (1954). It was supposedly based on some features of the structure of DNA, but these are irrelevant. The essential features of Gamow's scheme were as follows:

(a) Three bases coded one amino acid.
(b) Adjacent triplets of bases overlapped. See Fig. 1.
(c) More than one triplet of bases stood for a particular amino acid (degeneracy).

In other words it was an overlapping degenerate triplet code. Such a code imposes severe restrictions on the amino acid sequences it can produce. It is quite easy to disprove Gamow's code from a study of known sequences—even the sequences of the insulin molecule are sufficient. However, there are a very large number of codes of this general type. It might be thought almost impossible to disprove them all without enumerating them, but this has recently been done by Brenner (1957), using a neat argument. He has shown that the reliable amino acid sequences already known are enough to make *all* codes of this type impossible.

Attempts have been made to discover whether there are any obvious restrictions on the allowed amino acid sequences, although the sequence data available are very meagre (see the review by Gamow, Rich and Yčas, 1955). So far none has been found, and the present feeling is that it may well be that none exists, and that any sequence whatsoever can be produced. This is very far from being established, however, and for all we know there may be quite severe restrictions on the neighbours of the rarer amino acids, such as tryptophan.

If there is indeed a relatively simple code, then one of the most important biological constants is what Watson and I have called "the coding ratio" (Crick and Watson, 1956). If B consecutive bases are required to code A consecutive amino acids, the coding ratio is the number B/A, when B and A are large. Thus in Gamow's code its value is unity, since a string of 1000 bases, for example, could code 998 amino acids. (Notice that when the coding ratio is greater than unity stereochemical problems arise, since a polypeptide chain has a distance of only about 3½ Å. between its residues, which is about the minimum distance between successive bases in nucleic acid. However, it has been pointed out by

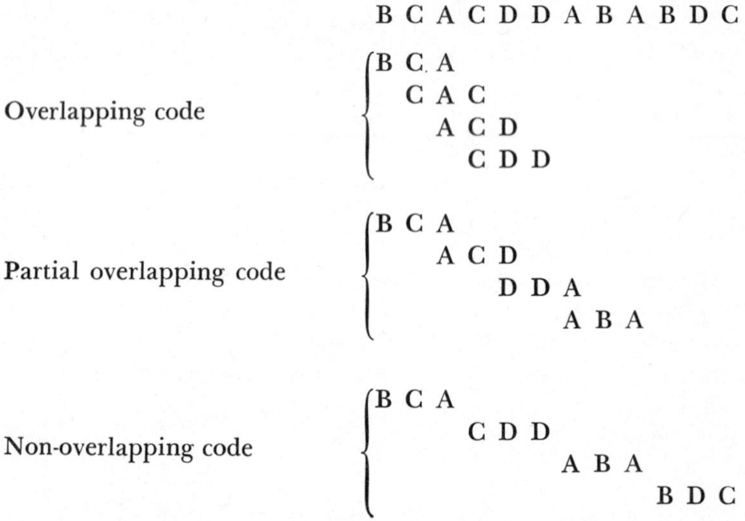

```
                        B C A C D D A B A B D C

                       {B C A
                       {  C A C
Overlapping code       {    A C D
                       {      C D D

                       {B C A
                       {    A C D
Partial overlapping code {        D D A
                       {            A B A

                       {B C A
                       {      C D D
Non-overlapping code   {            A B A
                       {                  B D C
```

FIG. 1. The letters A, B, C and D stand for the four bases of the four common nucleotides. The top row of letters represents an imaginary sequence of them. In the codes illustrated here each set of three letters represents an amino acid. The diagram shows how the first four amino acids of a sequence are coded in the three classes of codes.

Brenner [personal communication] that this difficulty may not be serious if the polypeptide chain leaves the template as it is being synthesized.)

If the code were of the non-overlapping type (see Fig. 1) one would still require a triplet of bases to code for each amino acid, since pairs of bases would only allow $4 \times 4 = 16$ permutations, though a possible but not very likely way round this has been suggested by Dounce, Morrison and Monty (1955). The use of triplets raises two difficulties. First, why are there not $4 \times 4 \times 4 = 64$ different amino acids? Second, how does one know which of the triplets to read (assuming that one doesn't start at an end)? For example, if the sequnce of bases is . . . , ABA, CDB, BCA, ACC, . . . , where A, B, C and D represent the four bases, and where ABA is supposed to code one amino acid, CDB another one, and so on, how could one read it correctly if the commas were removed?

Very recently Griffith, Orgel and I have suggested an answer to both these difficulties which is of some interest because it *predicts* that there should be only twenty kinds of amino acid in protein (Crick, Griffith and Orgel, 1957). Gamow and Yčas (1955) had previously put forward a code with this property, known as the "combination code" but the physical assumptions underlying their code lack plausibility. We assumed that some of the triplets (like ABA in the example above) correspond to an amino acid—make "sense" as we would say—and some (such as BAC and ACD, etc., above) do not so correspond, or as we would say, make "nonsense".

We asked ourselves how many amino acids we could code if we allowed all possible sequences of amino acids, and yet never accidentally got "sense" when reading the wrong triplets, that is those which included the imaginary commas. We proved that the upper limit is twenty, and moreover we could write down several codes which did in fact code twenty things. One such code of twenty triplets, written compactly, is

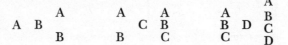

$$\mathrm{A\ B}\begin{matrix}\mathrm{A}\\ \\ \mathrm{B}\end{matrix}\qquad \begin{matrix}\mathrm{A}\\ \\ \mathrm{B}\end{matrix}\ \mathrm{C}\ \begin{matrix}\mathrm{A}\\ \mathrm{B}\\ \mathrm{C}\end{matrix}\qquad \begin{matrix}\mathrm{A}\\ \mathrm{B}\\ \mathrm{C}\end{matrix}\ \mathrm{D}\ \begin{matrix}\mathrm{A}\\ \mathrm{B}\\ \mathrm{C}\\ \mathrm{D}\end{matrix}$$

where $\mathrm{A\,B}\begin{matrix}\mathrm{A}\\ \mathrm{B}\end{matrix}$ means that two of the allowed triplets are ABA and

ABB, etc. The example given a little further back has been constructed using this code. You will see that ABA, CDB, BCA and ACC are among the allowed triplets, whereas the false overlapping ones in that example, such as BAC, ACD and DBB, etc., are not. The reader can easily satisfy himself that no sequence of these allowed triplets will ever give one of the allowed triplets in a false position. There are many possible mechanisms of protein synthesis for which this would be an advantage. One of them is described in our paper (Crick *et al.,* 1957).

Thus we have deduced the magic number, twenty, in an entirely natural way from the magic number four. Nevertheless, I must confess that I find it impossible to form any considered judgement of this idea. It may be complete nonsense, or it may be the heart of the matter. Only time will show.

V. CONCLUSIONS

I hope I have been able to persuade you that protein synthesis is a central problem for the whole of biology, and that it is in all probability closely related to gene action. What are one's overall impressions of the present state of the subject? Two things strike me particularly. First, the existence of general ideas covering wide aspects of the problem. It is remarkable that one can formulate principles such as the Sequence Hypothesis and the Central Dogma, which explain many striking facts and yet for which proof is completely lacking. This gap between theory and experiment is a great stimulus to the imagination. Second, the extremely active state of the subject experimentally both on the genetical side and the biochemical side. At the moment new and significant results are being reported every few months, and there seems to be no sign of work coming to a standstill because experimental techniques are inadequate. For both these reasons I shall be surprised if the main features of protein synthesis are not discovered within the next ten years.

It is a pleasure to thank Dr. Sydney Brenner, not only for many interesting discussions, but also for much help in redrafting this paper.

1957

REFERENCES

Beadle, G. M. (1945). *Chem., Rev. 37,* 15.

Benzer, S. (1957). In *The Chemical Basis of Heredity.* Ed. McElroy, W. D. and Glass, B. Baltimore: Johns Hopkins Press.

Berg, P. (1956). *J. Biol. Chem. 222,* 1925.

Borsook, H. (1956). *Proceedings of the Third International Congress of Biochemistry, Brussels,* 1955 (C. Liebecq, editor). New York: Academic Press.

Brachet, J. and Chantrenne, H. (1956). *Cold Spring Harb. Symp. Quant. Biol. 21,* 329.

Brenner, S. (1957). *Proc. Nat. Acad. Sci., Wash. 43,* 687.

Brown, N. H., Sanger, F., and Kitai, R. (1955). *Biochem. J. 60,* 556.

Cohen, G. N. and Cowie, D. B. (1957). *C. R. Acad. Sci., Paris, 244,* 680.

Cole, R. D., Coote, J., and Work, T. S. (1957). *Nature, Lond. 179,* 199.

Crick, F. H. C. (1957a). In *The Structure of Nucleic Acids and Their Role in Protein Synthesis,* p. 25. Cambridge University Press.

Crick, F. H. C. (1957b). In *Cellular Biology, Nucleic Acids and Viruses* 1957. New York Academy of Sciences.

Crick, F. H. C., Griffith, J. S., and Orgel, L. E. (1957). *Proc. Nat. Acad. Sci., Wash. 43,* 416.

Crick, F. H. C., and Watson, J. D. (1956). Ciba Foundation Symposium on *The Nature of Viruses.* London: Churchill.

Davie, E. W., Koningsberger, V. V., and Lipmann, F. (1956). *Arch. Biochem. Biophys. 65,* 21.

DeMoss, J. A., and Novelli, G. D. (1956). *Biochim. Biophys. Acta, 22,* 49.

Dounce, A. L. (1952). *Enzymologia,* 15, 251.

Dounce, A. L. (1956). *J. Cell. Comp. Physiol. 47,* suppl. 1, 103.

Dounce, A. L., Morrison, M., and Monty, K. J. (1955). *Nature, Lond. 176,* 597.

Fraenkel-Conrat, H. (1956). *J. Amer. Chem. Soc. 78,* 882.

Gale, E. F. (1953). *Advanc. Protein Chem. 8,* 283.

Gale, E. F. (1956). *Proceedings of the Third International Congress of Biochemistry, Brussels,* 1955 (C. Liebecq, editor). New York: Academic Press.

Gale, E. F., and Folkes, J. (1955). *Biochem. J. 59,* 661, 675 and 730.

Gamow, G. (1954). *Nature, Lond. 173,* 318.

Gamow, G., Rich, A., and Yčas, M. (1955). *Advanc. Biol. Med. Phys. 4*. New York: Academic Press.

Gamow, G., and Yčas, M. (1955). *Proc. Nat. Acad. Sci., Wash. 41,* 1101.

Gierer, A., and Schramm, G. (1956). *Z. Naturf. 11b,* 138; also *Nature, Lond. 177,* 702.

Goldstein, L., and Plaut, W. (1955). *Proc. Nat. Acad. Sci., Wash. 41,* 874.

Gros, F., and Gros, F. (1956). *Biochim. Biophys. Acta, 22,* 200.

Grunberg-Manago, M., and Ochoa, S. (1955). *J. Amer. Chem. Soc. 77,* 3165.

Harris, J. I., Sanger, F., and Naughton, M. A. (1956). *Arch. Biochem. Biophys. 65,* 427.

Hart, R. G., and Smith, J. D. (1956). *Nature, Lond. 178,* 739.

Hershey, A. D. (1956). *Advances in Virus Research.* IV. New York: Academic Press.

Hoagland, M. B. (1955). *Biochim. Biophys. Acta, 16,* 288.

Hoagland, M. B., Keller, E. B., and Zamecnik, P. C. (1956). *J. Biol. Chem. 218,* 345.

Hoagland, M. B., Zamecnik, P. C., and Stephenson, M. L. (1957). *Biochim. Biophys. Acta 24,* 215.

Hokin, L. E., and Hokin, M. R. (1954). *Biochim. Biophys. Acta, 13,* 401.

Hultin, T., and Beskow, G. (1956). *Exp. Cell Res. 11,* 664.

Ingram, V. M. (1956). *Nature, Lond. 178,* 792.

Ingram, V. M. (1957). *Nature, Lond. 180,* 326.

Kamin, H., and Handler, P. (1957). *Annu. Rev. Biochem. 26,* 419.

Keller, E. B., and Zamecnik, P. C. (1956). *J. Biol. Chem. 221,* 45.

Koningsberger, V. V., van de Ven, A. M., and Overbeck, J. Th. G. (1957). *Proc. K. Akad. Wet. Amst. B, 60,* 141.

Littlefield, J. W., and Keller, E. B. (1957). *J. Biol. Chem. 224,* 13.

Marmur, J., and Hotchkiss, R. D. (1955). *J. Biol. Chem. 214,* 383.

Mirsky, A. E., Osawa, S., and Allfrey, V. G. (1956). *Cold Spr. Harb. Symp. Quant. Biol. 21,* 49.

Munier, R., and Cohen, G. N. (1956). *Biochim. Biophys. Acta, 21,* 592.

Palade, G. E. (1955). *J. Biochem. Biophys. Cytol. 1,* 1.

Palade, G. E. (1956). *J. Biochem. Biophys. Cytol. 2,* 85.

Palade, G. E., and Siekevitz, P. (1956). *J. Biochem. Biophys. Cytol. 2,* 171.

Pardee, A. B. (1954). *Proc. Nat. Acad. Sci., Wash. 40,* 263.

Pardee, A. B., and Prestidge, L. S. (1956). *J. Bact. 71,* 677.

Pauling, L., Itano, H. A., Singer, S. J., and Wells, I. C. (1949). *Science, 110,* 543.

Petermann, M. L., and Hamilton, M. G. (1957). *J. Biol. Chem. 224,* 725; also *Fed. Proc. 16,* 232.

Petermann, M. L., Mizen, N. A., and Hamilton, M. G. (1952). *Cancer Res. 12,* 373.

Rabinovitz, M. and Olson, M. E. (1956). *Exp. Cell Res. 10,* 747.

Rabinovitz, M. and Olson, M. E. (1957). *Fed. Proc. 16,* 235.

Rich, A. and Davies, D. R. (1956). *J. Amer. Chem. Soc. 78,* 3548.

Schachman, H. K., Pardee, A. B., and Stanier, R. Y. (1953). *Arch. Biochem. Biophys. 43,* 381.

Schweet, R. (1957). *Fed. Proc. 16,* 244.

Simkin, J. L. and Work, T. S. (1957a). *Biochem. J. 65,* 307.

Simkin, J. L. and Work, T. S. (1957b). *Nature, Lond. 179,* 1214.

Simpson, M. V. and McLean, J. R. (1955). *Biochim. Biophys. Acta, 18,* 573.

Simpson, M. V., McLean, J. R., Cohn, G. I., and Brandt, I. K. (1957). *Fed. Proc. 16,* 249.

Spiegelman, S. (1957). In *The Chemical Basis of Heredity.* Ed. McElroy, W. D. and Glass, B. Baltimore: Johns Hopkins Press.

Steinberg, D. and Mihalyi, E. (1957). In *Annu. Rev. Biochem. 26,* 373.

Steinberg, D., Vaughan, M., and Anfinsen, C. B. (1956). *Science, 124,* 389.

Stephenson, M. L., Thimann, K. V., and Zamecnik, P. C. (1956). *Arch. Biochem. Biophys. 65,* 194.

Synge, R. (1957). In *The Origin of Life on the Earth.* U.S.S.R. Acad. of Sciences.

Ts'o, P. O. B., Bonner, J., and Vinograd, J. (1956). *J. Biochem. Biophys. Cytol. 2,* 451.

Wagner, R. P. and Mitchell, H. K. (1955). *Genetics and Metabolism.* New York: John Wiley and Sons.

Yčas, M. and Brawerman, G. (1957). *Arch. Biochem. Biophys. 68,* 118.

Zamecnik, P. C., Keller, E. B., and Littlefield, J. W., Hoagland, M. B. and Loftfield, R. B. (1956). *J. Cell. Comp. Physiol. 47,* suppl. 1, 81.

Ernst Freese

On the Molecular Explanation of Spontaneous and Induced Mutations

Muller showed that mutations could be induced by X rays. His analysis of mutation over the years revealed that the majority of spontaneous mutations in the fly, and probably for man, were of a chemical nature rather a consequence of background radiation. He also found that many mutations are point mutations, presumably alterations of a few atoms within a complex molecule. Benzer produced a fine-structure map of spontaneous mutation in bacteriophage. Ernst Freese used chemical mutagens, some of which were related to nucleotides. These substituted nucleotides or base analogues caused mutations whose sites of damage did not correspond to the sites produced by most of the spontaneous mutations and not at all to the sites produced by the acridine dyes. The mechanism of *transitions* proposed by Freese is accepted as one mechanism by which mutation occurs. The hypothesis of "transversions" proposed for acridines by Freese has been replaced by a more complex hypothesis formulated by Crick and his colleagues. The transition substitutions obtained by Freese can also be demonstrated *in vitro* with DNA molecules.

INTRODUCTION

For a molecular understanding of heredity it is necessary to know the molecular structure of the "genome" in which most hereditary determinants, or "genes," are one-dimensionally arranged. Fortunately, many recent observations indicate that the hereditary information is determined by DNA, whose molecular structure is sufficiently known,[1] while other molecules, including proteins, keep the DNA molecules together and in the right sequence. This

Reprinted from *Brookhaven Symposia in Biology 12*:63–73, 1959, by permission of the publisher and the author.

is, at least, an excellent working hypothesis. It implies that "genes" are correlated, in some way, to DNA. This correlation can be detected by comparing genetic observations in very *small* dimensions with the molecular properties of DNA. A molecular understanding of *large* chromosomal changes, however, must await the chemical analysis of the connections that keep DNA molecules together.[2] The following remarks will therefore be limited to the analysis of genetic fine structure due to the molecular properties of DNA.

It seems clear that the information contained in DNA must be specified by the sequence of nucleotide pairs along the molecule. How much of this information is redundant, and can be changed arbitrarily without altering the genetic expression, is not yet known. But the large number of specific enzymes and other molecules in a cell demands quite a large number of nucleotide pairs to determine the specific property of any given enzyme or constitutive protein. The specificity of such a protein must be altered when only a small number, perhaps only one, of these necessary nucleotide pairs is changed by a "mutation."

This expectation is in excellent agreement with many recent genetic fine structure observations, most of which are thus far limited to microorganisms. The most detailed examinations have been done by Benzer[3,4] for the *r*II mutants of phage T4.

All these observations agree with the notion that a "gene" corresponds to a number of nucleotide pairs. If a gene is to be defined at all in molecular terms, it may be identified with a unit of function which is determined by a stretch of many nucleotide pairs of the order of 10^3. It is then automatically impossible, however, to call this gene a unit of recombination or of mutation, for *recombination* does occur within a functional unit even if mutations are only a few nucleotide pairs distant, and a *mutation* may involve any portion of a functional unit. Thus, the classical concept of a gene, which is useful to describe genetic observations in *large dimensions,* has to be replaced by a more sophisticated notion of nucleotide sequences in *molecular dimensions.*

In this paper an attempt will be made to show that the change of a single nucleotide pair in DNA is sufficient to cause a mutation, and the way in which such a change may come about will be derived. Ideally, one would like to determine the mutagenic changes directly, by chemically comparing the nucleotide sequence of standard type and mutant DNA. This is impossible, however,

at least at present. Instead, one must proceed more indirectly and derive a molecular picture by comparing genetic and chemical observations on nucleic acids.

THE EXTENT OF MUTATIONS

The genetic analysis[3] of many spontaneous mutations of phage T4 has shown that most of them are localized in a very small area of the genome. To find out how many nucleotide pairs have to be changed in order to cause such a "point mutation," mutations must be induced artificially, by agents which react with DNA in a known manner, or at least in a way amenable to chemical analysis. This will indicate which initial molecular reaction leads to such an induced mutation. One may then hope to understand spontaneous mutations by comparing their genetic properties with those of the induced ones.

For most of our studies, we have used two classes of mutagenic agents.

1) The first class comprises close *analogues* of normal nucleic acid bases or nucleosides which are known to be, or can be expected to be, incorporated into DNA. The specific enzymes of nucleotide metabolism must be able to convert these base analogues into direct DNA precursors (deoxynucleoside triphosphates), which then must be incorporated into replicating DNA. Nucleotides as such do not seem to be taken up by a cell.

At present, three highly mutagenic base analogues of this kind are known, by which 1% *r*-type mutants are readily induced in multiplying phage T4: 5-bromouracil (BU) [or 5-bromodeoxyuridine (BD)],[5] 2-aminopurine (AP),[6] and 2,6-diaminopurine.[6] BU can quantitatively replace thymine in bacterial[7] and phage[8] DNA. The incorporation of AP is structurally probable,[6] although experiments show it to be $< 1\%$ that of adenine, if it takes place at all.[9]

2) The second class consists of agents which alter resting DNA in such a way that mutations result in subsequent DNA replications. We are not interested here in scissions of DNA molecules, but in small changes of only one or a few nucleotides. Such changes can be obtained, e.g., by the slight alteration of some bases, or by their removal.

a) For the *alteration of nucleic acid bases,* many chemical reactions are obviously feasible, but the most interesting are those

that specifically alter the hydrogen-bonding properties towards the complementary bases in DNA. Nitrous acid, e.g., causes the replacement of amino groups by hydroxl groups, which, by a tautomeric shift, probably change to keto groups in the bases of interest here.

Nitrous acid has been found to be highly lethal[10] and mutagenic[11] for the RNA-containing tobacco mosaic virus (TMV), for the single stranded DNA phage ΦX174,[12] and for the double stranded DNA phage T4.[13,14] Figure 1 and Table I show the conditions under which mutants were isolated[14] from *r* plaques, mostly mottled.

b) For the removal of bases, the treatment of DNA by low *p*H would seem most adequate, since this is known to remove purines rapidly.[15] We[14] have treated T4 phages at *p*H 5, at which protein denaturation is not very extensive, and increased the speed of the reaction by running it at 45°C. At a survival of 4×10^{-5} we found in phage T4 0.36% *r* plaques, mostly mottled, and isolated the mutants (see Figure 2 and Table I). Zamenhof and Greer[16] re-

TABLE I

FREQUENCY OF *r*-TYPE MUTANTS AFTER PHAGE TREATMENT *In Vitro*

Treatment	*p*H	Treatment time, hr.	*r* + Mottled plaques $\times 10^{-3}$	Total *r*II mutants $\times 10^{-3}$
None	7.4	—	0.36	0.21
Nitrous acid* at 25°C	5.3	1/6	3.0	1.4
Acetate buffer** at 45°C	4.5	7	2.5	†
	4.7	7	1.4	†
	5.0	27	1.7	†
	5.0	43	3.6	1.6

* See Figure 1 for details of treatment.
** See Figure 2 for details of treatment.
† Not tested.

port that, even at neutral *p*H, high temperatures were mutagenic for both wet and dry bacteria; they[17] have furthermore observed the liberation of small amounts of purines from DNA under such conditions.

It can be assumed that all the agents just described are mutagenic because of their direct action upon nucleic acid. This is plausible for the direct chemical action on *free* T4 phage, since, at infection, most of the phage protein stays outside the bacterium. In the case of TMV, Mundry and Gierer[11] have proven this view directly. They find the same percentage of mutants in TMV particles treated with nitrous acids as in the RNA isolated from them. For base analogue induced mutants, a proof of the direct mutagenic action has not yet been given, but the ability of these bases to pair with the normal DNA bases renders this assumption likely.

Because of the chemical nature of the mutagens, it can further be assumed that the *first change* in nucleic acid leading to a mutation involves only *one* of the nucleic acid bases. This is supported by the observation on TMV[11] and phage T2[13b] that the ratio of induced mutants to viable virus increases linearly with the exposure time to nitrous acid.

Although the first mutagenic change in nucleic acid seems to involve only one base, the ultimate mutation could extend over more than one base, or base pair, if further changes are caused by the first one in the subsequent replication. Indeed, 7% of the heat induced mutants in phage T4 do not revert and often show a larger genetic alteration; they may be caused by such a mechanism.

However, most of the temperature induced mutants and practically all of the mutants induced in all other ways do revert spontaneously, and their mutations are confined to a very small area of the genetic map.[6, 14, 18] It is very probable, therefore, that most of these induced "point mutations" are due to the change of a single nucleotide pair in phage T4 DNA. Since most spontaneous mutations can also revert spontaneously and are genetically well localized, it seems likely that they too arise by the change of a single nucleotide pair.

MUTAGENIC SELECTION RULES

The next question is which base or base pair change of nucleic acid concurs with a given point mutation, and how this change

comes about. It is clear that not all possible changes will be induced with equal frequency. On the contrary, it may be expected that certain base or base pair changes cannot be induced at all by a given mutagen. In other words, one may expect to find selection rules which govern the induction of different mutations.

One such selection rule for the induction of forward and reverse mutations in double-stranded DNA will be discussed. The experiments were performed with spontaneous and induced *r*II-type mutants of phage T4, and are described in more detail elsewhere.[19] The *r*II mutants are advantageous for our reversion studies, since no suppressor mutations have been observed at other loci. Actually, most spontaneous or induced revertants of the standard phenotype probably arise by genuine back mutations to the standard genotype.[19]

We have examined many spontaneous and induced *r*II-type mutants for the induction of reverse mutations by the two base analogues, AP and BD. This can be done by a simple spot test.[19] The results (Table II) show that mutants induced by BU, BD, or AP can also be induced to revert by these base analogues. Similarly, most mutants induced by nitrous acid, or by lower *p*H and higher temperature, can be induced by the base analogues to revert. The noninducible mutants may all belong to the background of spontaneous mutants, most of which are noninducible (see Table II). This is also indicated by the genetic fine structure analysis; most of the non-reversion inducible mutants belong to genetic sites for which some spontaneous mutants have been found in the standard type phage stock from which they were derived. At most, a small percentage of the mutants induced by lower *p*H and higher temperature may not be inducible to revert by the base analogues.

In contrast to these base analogue inducible mutants, most of the spontaneous, and practically all of the proflavine induced mutants, could not be induced to revert by either base analogue, although they readily revert spontaneously. (Proflavine induction will be discussed below.) These mutants do not seem to be inducible at all by our base analogues, since induction was easily observable for the inducible mutants, whether the background of spontaneous revertants was high or low.

We have thus found a mutagenic selection rule which permits us to distinguish between two different mutagenic effects. The

next step is to correlate these observations with the different changes of nucleotide pairs in DNA. First, however, an assumption must be made about DNA replication; for, each original change of a DNA base which causes a mutation leads to a complete change of a nucleotide pair only after some DNA replications. We postulate that the nucleotide sequence in a new DNA chain is determined in each portion by complementary base pairing with one pre-existing DNA chain, which need not be the same chain for the whole new DNA molecule. This assumption includes a number of possible replication schemes (e.g., see reference 1) and is supported both by the results of Meselson and Stahl[20] concerning bacterial DNA replication and by the following observations about mutations: Under any of the mutagenic treatments mentioned before, many hybrid T4 phage particles are obtained which give rise to mottled *r*-plaques. This observation, which recently has been examined quantitatively,[12, 21] in-

TABLE II

REVERSION INDUCTION OF REVERTING *r*II-TYPE MUTANTS

Mutagen	Mutants tested	Base analogue inducible, %	Base analogue non-inducible, %	Background of spontaneous non-inducible mutants, %
2-Aminopurine[6]	98	98	2	2
5-Bromouracil[18]	64	95	5	2
Nitrous acid*	47	87	13	15
*p*H 5, 45°C*	115	77	23	15
Spontaneous[3, 18]	110	14	86	..
Proflavine[22]	55	2	98	..

* Described in this paper.
Reversion induction is described in more detail in reference 19.

dicates that both the normal and the altered chain of DNA can serve separately as templates for new DNA molecules.

The next question is which base pair changes can be induced by the different mutagens.

1) The large mutagenic effect of our *base analogues* is probably due to mistakes in complementary base pairing which accompany their incorporation into DNA. This has been discussed in detail elsewhere.[6] In each pairing mistake, one purine is replaced by another purine, or one pyrimidine by another pyrimidine; i.e., in each single DNA chain, the number of purines, as well as of pyrimidines, is conserved at any stage of the mutagenic process. After sufficient DNA replications in a medium no longer containing the base analogue, mutant DNA will be present in which one base pair is replaced by another one. Because of the purine-pyrimidine conservation, the only possible base pair changes that can be induced by such base analogues are the "transitions":

$$\begin{matrix} A & & G \\ & \longleftrightarrow & \\ T & & C \end{matrix} \qquad \text{(I)}$$

(A = adenine, G = guanine, T = thymine, and C = cytosine, or any derivative which is altered at the 5-position, e.g., 5-methylcytosine, 5-hydroxymethylcytosine, etc.)

Furthermore, each base analogue should be able to induce transitions (I) in both directions: in one direction by a pairing mistake made by the base analogue when it becomes incorporated into DNA; in the other direction by a pairing mistake which the base analogue makes in later replications, after it has been incorporated into DNA.[6]

In other words, it would be expected that each base analogue induced mutation can also be induced to back mutate by the same base analogue, or by any other one that induces transitions (I). This expectation agrees very well with our observations (Table II).

2) The induction of point mutations by *nitrous acid* is almost certainly due to the direct deamination of DNA bases. Some of the new bases have altered pairing properties and cause the change of a nucleotide pair in subsequent DNA replication. Thus, nitrous acid alters:

A into hypoxanthine, with pairing properties like those of G, and should cause the ultimate change

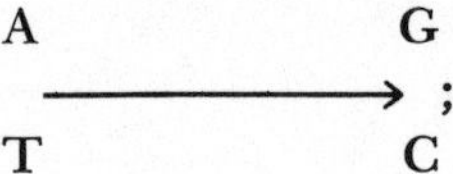

C into uracil, with pairing properties like those of T, and should cause the ultimate change

G into xanthine, with pairing properties like those of G, and should cause no base pair change.

Again, the number of both purines and pyrimidines should be conserved in each DNA chain, and the transitions (I) might be induced in both directions. This agrees well with our observation that nitrous acid induced mutants are inducible to revert by our base analogues. They might also be reversion inducible by nitrous acid itself, but this has not yet been tested.

3) Rather interesting is the mutagenic effect of lower *p*H and higher temperatures, which is very likely due to the separation of single purines from the sugar phosphate backbone. Originally, it was not even clear whether such a base removal would prevent the DNA from replicating across the gap in its structure and either be lethal or induce only large alterations in the chromosome. However, some repair mechanism seems to work, at least in some cases, since we observed the induction of point mutations under such a treatment. Either the removed purine gets replaced by another (activated) base in the *resting* DNA or during DNA *replication* the new DNA chain, using the altered chain as the complementary template, incorporates a base at the site of the gap which it would not have incorporated normally. If any of the four bases entered this site more or less at random, one would expect the following:

a) When the same base enters the DNA chain that is contained in it normally, no mutation results.

b) When one purine replaces the other purine or one pyrimidine the other one, a transition (I) results.

c) When a purine is replaced by a pyrimidine or vice versa a new base pair change is induced such that in some progeny DNA molecules an original pyrimidine is replaced by a purine and vice versa. In this way, any one of the following "transversions" of base could be induced:

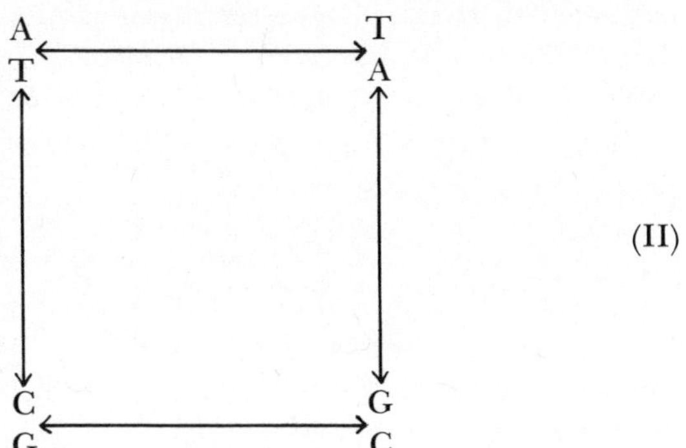

Although the *sum* of purines plus pyrimidines per DNA chain is again conserved for these base pair changes, the *individual* number of purines or pyrimidines is not. On the contrary, for each transversion, the number of purines increases and the number of pyrimidines decreases by one, or vice versa, in a given DNA chain. None of these transversions should be inducible either by base analogues or by nitrous acid. In other words, such mutations should not be inducible by our base analogues to back mutate.

Surprisingly, we have found that most of the mutants induced by lower *p*H and higher temperature *can* be induced to revert by our base analogues; i.e., most of the induced mutants seem to be due to transitions (I). More detailed analyses are necessary before one can decide between several possible explanations.

4) Although we did not observe any appreciable induction of non-base analogue inducible mutants by heat, mutagenic transversions (II) apparently do exist. For, about 86% of the examined spontaneous mutants could not be induced to revert by either base analogue, although they revert spontaneously. They probably contain a transversion (II). However, 14% of the spontaneous mutants could be induced to revert by base analogues and are therefore mostly due to transitions (I). Hence, both base pair changes (I) and (II) can arise spontaneously, although with different frequency. This explains why all point mutations can revert spontaneously.

The existence of two different spontaneous mutagenic effects is also shown by the reversion properties of some base analogue induced mutants.[19] These mutants throw off both partial revertants (tiny plaques) and revertants of standard phenotype. After growth in the presence of base analogues, however, the

frequency of standard type revertants is highly increased, while that of the partial revertants is unaltered.

The origin of spontaneous mutations is not known. However, since they arise during phage growth (the mutant frequency of a phage stock kept cold does not alter appreciably), it is likely that they are caused during DNA replication by the mistaken incorporation of a wrong base: for transversions (II), the mistaken pairing of two normal purines, or two normal pyrimidines, may be responsible,[19] while transitions (I) may arise by the replacement of one purine by another purine (a normal one[1] or a base analogue), or one pyrimidine by another. It is not yet clear whether all four transversions (II) can arise spontaneously or only some of them.

5) Proflavine has not been mentioned so far. Its mutagenic action can be explained by several molecular reactions on *replicating* DNA.[19, 22] Which of the possible transversions (II) are actually induced is not yet known. But whatever the precise mechanism may be, one would expect that these mutants should be reversion inducible by proflavine itself.

MUTATION PROBABILITIES

So far two mutually exclusive mutagenic effects have been discussed. The next question is, what is the *probability* of a certain base pair change occurring anywhere in DNA under a given mutagenic treatment. There is no simple answer, since large differences in mutation frequencies have been observed.

1) The spontaneous reversion indices vary greatly[3] from mutant to mutant, over several powers of 10. This should be due, in part, to the different spontaneous mutagenic mechanisms. But even among the base analogue inducible or noninducible mutants alone, large differences in reversion indices have been observed which cannot yet be grouped into simple classes.

2) Base analogue inducible mutants vary in their response to reversion induction by AP and BD. Some are highly inducible by AP and little by BD, others vice versa, and some show about the same response with both base analogues.[19]

3) The genetic fine structure analysis of many independently isolated *r*II-type mutants has shown that spontaneously,[4] as well as with BU[18] and AP[6] induction, mutations are preferentially induced at certain genetic sites ("hot spots"). We[14] have also de-

termined the relative genetic position of nitrous acid and heat induced mutations, and compared them to the background of spontaneous mutations present in the phage stock used. The results are shown in Table III. Neither the nitrous acid nor the heat induced mutants exhibit pronounced sites of high mutability, in contrast to the mutants previously described, all of which arose during phage growth.

All these observations about the various mutagenic specificities show that the mutation probability depends both on the kind of base pair itself and on the position of this base pair within DNA. This may be due to at least three different causes:

a) A difference in the relative frequency of direct DNA precursors. This would only distinguish between the two kinds of base pairs and not cause a very specific mutagenic preference.

b) An effect of neighboring bases upon the mutation properties of a given base (e.g., neighboring A-T pairs may facilitate the separation of a given pair). This effect may be sufficient to explain the deviation from a Poisson distribution for the mutability spectra of chemically treated free phages. Whether hot spots of mutation can also be explained by this effect alone is not yet clear.

c) A specific protein. For example, the DNA synthesizing enzyme (or enzymes) may exert a preference both towards certain nucleotides in the DNA precursor pool and towards certain base configurations in the DNA template. Possibly, the highly specific hot spots of mutation are due to such a protein specificity. In any case, the absence of hot spots for the mutations of chemically treated free phages suggests that this highly specific mutagenic effect occurs only inside the bacterium during DNA replication.

CONCLUSION

Reasons have been offered for the view that most reverting point mutations of phage T4 correspond to the change of a single nucleotide pair in DNA. Furthermore, the existence has been shown of a mutagenic selection rule which permits the division of point mutations into two classes. One class corresponds to base pair changes for which the number of purines as well as of pyrimidines is conserved in each DNA chain. The other class seems to comprise all those base pair changes for which a purine is exchanged for a pyrimidine and vice versa; this class should be further sub-

TABLE III

NUMBERS OF GENETIC SITES AT WHICH n MUTATIONS HAVE BEEN OBSERVED, FOR INDEPENDENTLY ISOLATED r-II TYPE MUTANTS

	n																					
	1	2	3	4	5	6	7	8	9	10	11	12	13	14	15	16	17	18	19	20	24	27
Spontaneous[7]	31	13	5	1		1			1		1								1			
BU+BD[7]	12	3	6	2	2	2			1					1			1			1		1
				[1]		[1]																
AP[7]	28	7	2	3				1	1												1	
*p*H 5, 45°C*	33	18	4	2	1	4																
		[4]	[1]																			
Nitrous acid*	17	8	2	2																		
		[1]																				
Example of Poisson distribution	32	19	8	2	1	0	0	0	0													

* Described in this paper.

The numbers in brackets are sites at which several spontaneous mutants were observed in the original standard type stock; these sites are included in the main figures but they probably belong to the background of spontaneous mutants. The example of a Poisson distribution has been calculated for an average number of 1.2 mutants per site. The number of sites with no mutant would then be 27.

divisible by more selection rules when suitable mutagenic agents are used, unless DNA confers its information by a binary code.[23] Finally, different possible causes for the observed higher mutagenic specificity have been discussed.

Although the experiments have been limited to viruses so far, I have no doubt that most point mutations must be governed by the same rules in other organisms. For, it seems clear that the mutagenic characteristics of phage T4 are due to the molecular properties of *all* double stranded DNA molecules and are not limited to the specific characteristics of a particular phage T4 DNA molecule.

1959

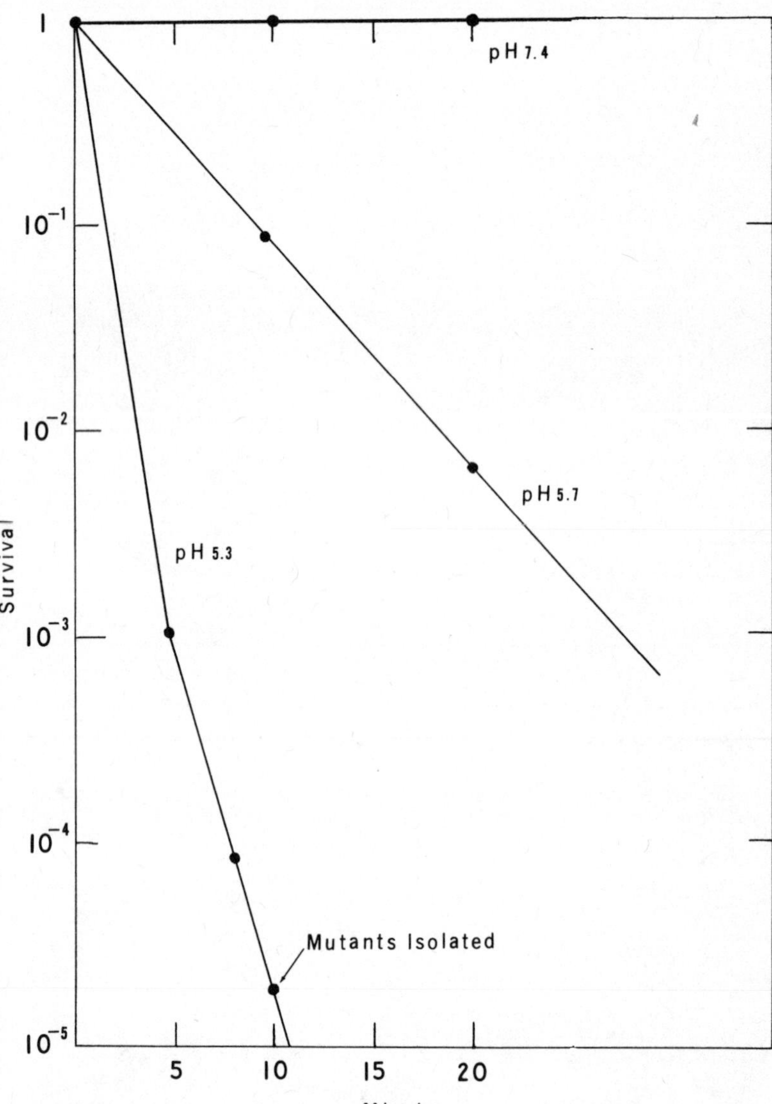

FIG. 1. Inactivation of T4 in nitrous acid. T4 standard type phages were diluted 1:50 or 1:100 in 2 *M* nitrous acid + 10^{-3} *M* $MgSO_4$, adjusted to the proper *p*H by 1 *N* HCl. The reaction took place at 25°C and was stopped by 10-fold dilution of 0.75 *M* phosphate buffer + 10^{-3} *M* Mg at *p*H 7.4.

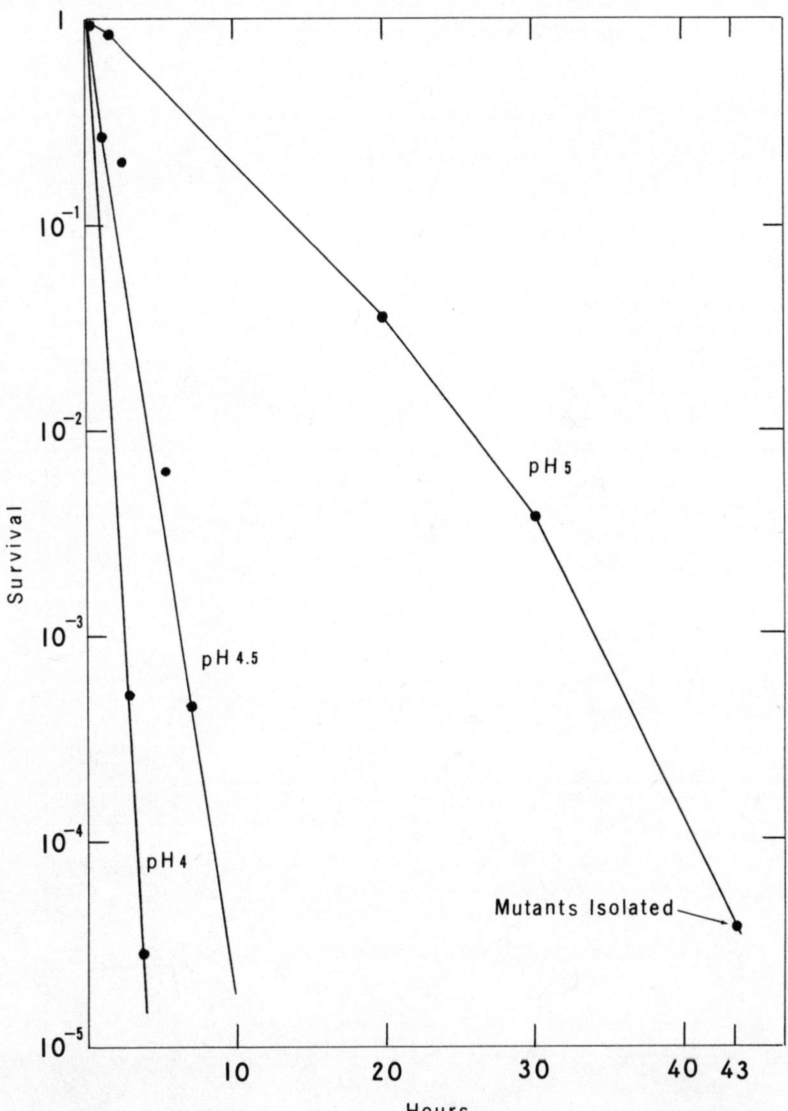

FIG. 2. Inactivation of T4 by lower *p*H and 45°C. T4 standard type phages were diluted 1:70 in 0.1 *M* Na acetate + 10^{-3} *M* $MgSO_4$, adjusted to the proper *p*H by acetic acid. The reaction took place at 45°C and was stopped by 10-fold dilution in tryptone broth. At 37°C, both inactivation and mutant induction were slower.

REFERENCES

[1] Watson, J. D. and Crick, F. H. C., *Cold Spring Harbor Symposia Quant. Biol.* 18, 123 (1953).

[2] Freese, E., *Cold Spring Harbor Symposia Quant. Biol.* 23, 13 (1958).

[3] Benzer, S., *Proc. Natl. Acad. Sci. U.S.* 41, 344 (1955).

[4] Benzer, S., in *The Chemical Basis of Heredity,* McElroy and Glass, Editors, Johns Hopkins Press, Baltimore, Md., 1957.

[5] Litman, R. M., and Pardee, A. B., *Nature* 178, 529 (1956).

[6] Freese, E., *J. Mol. Biol.* 1 (1959), in press.

[7] Zamenhof, S., and Griboff, G., *Nature* 174, 306 (1954).

[8] Dunn, D. B. and Smith, J. D., *Nature* 174, 304 (1954).

[9] Freese, E., Unpublished data.

[10] Schuster, H. and Schramm, G., *Z. Naturforsch.* 13b, 697 (1958).

[11] Mundry, K. W. and Gierer, A., *Z. Vererbungsl.* 89, 614 (1958).

[12] Tessman, I., Personal communication.

[13a] Benzer, S., Personal communication.

[13b] Vielmetter, W. and Wieder, C. M., *Z. Naturforsch.* 14b, 312 (1959).

[14] Freese, E. and Freese-Bautz, E., Data in this paper.

[15] E.g., Chargaff, E., *The Nucleic Acids,* Academic Press, New York, 1955.

[16] Zamenhof, S. and Greer, S., *Nature* 182, 611 (1958).

[17] Greer, S. and Zamenhof, S., *Federation Proc.* 18, 938 (1959).

[18] Benzer, S., and Freese, E., *Proc. Natl. Acad. Sci. U.S.* 44, 112 (1958).

[19] Freese, E., *Proc. Natl. Acad. Sci. U.S.* 45, 622 (1959).

[20] Meselson, M. and Stahl, F. W., *Proc. Natl. Acad. Sci. U.S.* 44, 671 (1958).

[21] Pratt, D. and Stent, G. S., *Proc. Natl. Acad. Sci. U.S.,* in press.

[22] Brenner, S., Benzer, S., and Barnett, L., *Nature* 182, 983 (1958).

[23] Crick, F. H. C., See paper in this symposium.

Marshall W. Nirenberg and Heinrich Matthaei

The Dependence of Cell-Free Protein Synthesis in *E. Coli* upon Naturally Occurring or Synthetic Template RNA

In 1959 most geneticists and molecular biologists believed that a solution to the coding problem was imminent. Several teams were in competition and all were using the same conceptual basis. The principle of the colinearity of the protein's amino-acid sequence, the DNA nucleotide sequence, and the genetic map provided the basis for their approach. They used a selected protein (bacteriophage lysozyme or head protein) and they employed Ingram's chromatographic techniques to the mutants affecting these proteins. They also used specific mutagens (e.g., base analogues; acridines) in the hope that correspondences could be made between fine structure maps and the corresponding protein "fingerprints." To everyone's surprise, they were all scooped by Marshall Nirenberg and his colleagues. Nirenberg used a protein-synthesizing system which required a special RNA—now known as messenger RNA because it carries the genetic code from the nuclear DNA to cytoplasmic membranes where protein synthesis takes place. As a messenger RNA, Nirenberg used a simple polymer of uracil. When amino-acid incorporation occurred, Nirenberg found that the polyuracil was decoded as polyphenylalanine. Thus the first coding sequence UUUUUUUUU . . . could be translated as phe.phe.phe. . . . Crick designated the minimum coding sequence as a *codon*. The codon contains a sequence of *three* nucleotides and it codes *one* amino acid. The decoding of the nucleotide sequences has resulted in a "syllabary" for the sixty-four combinations of codons for the four

Reprinted from *Proceedings of the Fifth International Congress of Biochemistry,* Moscow, 1961 (New York: Pergamon Press, Inc., 1963), pp. 184–189, by permission of the publisher and the authors.

nucleotides of DNA. Almost all combinations code. There is also evidence for the universality of the code. The protein-synthesizing machinery of a sea-urchin egg or a bacterium can decode poly-uracil as polyphenylalanine.

We have obtained a stable, cell-free *E. coli* system which incorporates ^{14}C-L-valine into protein at a rapid rate and which has many characteristics of protein synthesis. Conditions have been found which demonstrate a novel characteristic of this system; that is, a requirement for ribosomal RNA, needed even in the presence of excess soluble RNA and ribosomes. Naturally occurring RNA, as well as a synthetic polynucleotide, were active in this system. The synthetic polynucleotide appears to contain the code for the synthesis of a protein containing only one amino acid.

In Fig. 1, radioactive counts per minute per milligram protein is plotted against time in minutes. In the absence of added DNAase, valine was rapidly incorporated into protein. At the end of 90 min the extracts were still active and 2000 counts had been incorporated. The incorporation required ribosomes, $100{,}000 \times g$ supernatant solution and ATP and an ATP-generating system. The incorporation was markedly inhibited by RNAase, puromycin and chloramphenicol. Addition of 10 μg of crystalline DNAase/ml of reaction mixture did not affect the initial rate of incorporation, but stopped the incorporation after 30 min. The system was extremely sensitive to DNAase; as little as 0.1 μg was inhibitory.

Addition of ribosomal RNA to the reaction mixtures increased the amount of valine incorporated. Soluble RNA did not replace ribosomal RNA. In most of the work that will follow, RNA was added after 30 min of incubation, and amino acid incorporation then was almost completely dependent upon the addition of ribosomal RNA.

In Fig. 2, ^{14}C-L-valine incorporation is presented as a function of time. Addition of ribosomal RNA markedly increased incorporation. Total incorporation was proportional to the amount of ribosomal RNA added, suggesting a stoichiometric rather than a catalytic action of ribosomal RNA. In contrast, Hoagland and Comly have recently shown S-RNA to act in a catalytic fashion. Preliminary fractionation of ribosomal RNA indicated that only a small portion of the RNA was active; most of the ribosomal RNA seemed to be inactive.

TABLE I

POLYNUCLEOTIDE SPECIFICITY FOR PHENYLALANINE INCORPORATION

Expt. No.	Additions	Counts/min/ mg protein
1	None	44
	+10 μg Polyuridylic acid	39,800
	+10 μg Polyadenylic acid	50
	+10 μg Polycytidylic acid	38
	+10 μg Polyinosinic acid	57
	+10 μg Polyadenylic-uridylic acid (2/1 ratio)	53
	+10 μg Polyuridylic acid + 20 μg polyadenylic acid	60
	Deproteinized at zero time	17
2	None	75
	+10 μg UMP	81
	+10 μg UDP	77
	+10 μg UTP	72
	Deproteinized at zero time	6

A number of synthetic polynucleotides were tested to see if they would replace ribosomal RNA in stimulating amino acid incorporation (Table I). 10 μg polyuridylic acid per ml. of reaction mixture greatly stimulated the incorporation of phenylalanine. The stimulation was highly specific, for other synthetic polynucleotides were not active. The absolute specificity of poly-U was confirmed by demonstrating that a randomly mixed polymer containing adenylic and uridylic acids at a 2:1 ratio was inactive. A solution of poly-U and poly-A, which base pairs to form double- and triple-stranded helices, had no activity whatsoever, suggesting that single-strandedness is a necessary requisite for activity. Experiment 2 demonstrates that UMP, UDP, and UTP were unable to stimulate phenylalanine incorporation.

In Fig. 3 counts per minute of U-^{14}C-L-phenylalanine incorporated per milligram protein is plotted against time in minutes. No incorporation occurred in the absence of poly-U. When 10 μg poly-U per ml. reaction mixture were added, 70 per cent of

TABLE II

CHARACTERISTICS OF POLYURIDYLIC ACID DEPENDENT PHENYLALANINE INCORPORATION

Additions	Counts/min/ mg protein
Minus polyuridylic acid	70
None	29,500
Minus 100,000 × *g* supernatant solution	106
Minus ribosomes	52
Minus ATP, PEP and PEP kinase	83
+ 0.02 μmoles puromycin	7100
+ 0.31 μmoles chloramphenicol	12,550
+ 6 μg RNAase	120
+ 6 μg DNAase	27,600
Minus amino acid mixture	31,700
Deproteinized at zero time	30

the 0.07 μM of U-^{14}C-L-phenylalanine present was incorporated.

The data of Table II show the characteristics of phenylalanine incorporation. Little incorporation was observed in the absence of poly-U. 10 μg poly-U were present in all other samples. The large phenylalanine incorporation due to poly-U, required both 100,000 × *g* supernatant solution and ribosomes; was dependent upon ATP and an ATP-generating system; and was inhibited by puromycin, chloramphenicol and RNAase, but not DNAase. Omitting a mixture of 30 amino acids had no effect upon phenylalanine incorporation, suggesting the synthesis of a polypeptide or protein containing phenylalanine alone. This conclusion was substantiated by demonstrating that poly-U did not appreciably stimulate the incorporation of any other ^{14}C-amino acid.

The reaction product obtained after phenylalanine had been incorporated was partially characterized and the results are presented in Table III. The characteristics of the product of the reaction and authentic poly-L-phenylalanine are shown. Unlike many other polypeptides and proteins, both the product of the reaction and the polymer were only partially hydrolyzed by 6 N HCl and were hydrolyzed completely by 12 N HCl. Polyphenylalanine was soluble in 33 per cent HBr in glacial acetic

acid and the product of the reaction was also. Both polyphenylalanine and the product were insoluble in many other solvents. The product of the reaction could be purified by means of its unusual solubility behavior. Seventy percent of the total amount of ^{14}C-phenylalanine incorporated into protein due to the addition of poly-U could be recovered by extraction with HBr in glacial acetic acid and subsequent precipitation. Complete hydrolysis of the purified product followed by paper electrophoresis demonstrated that the isotope was in the form of phenylalanine. No other radioactive spots were found.

TABLE III

COMPARISON OF CHARACTERISTICS OF PRODUCT OF REACTION AND POLY-L-PHENYLALANINE

Treatment	Product of reaction	Poly-L-phenylalanine
6 N HCl for 8 hr at 100°	Partially hydrolyzed	Partially hydrolyzed
12 N HCl for 48 hr at 120–130°	Completely hydrolyzed	Completely hydrolyzed
Extraction with 33% HBr in glacial acetic acid	Soluble	Soluble
Extraction* with the following solvents: H_2O, benzene, nitrobenzene, chloroform, N, N-dimethyl-formamide, ethanol, petroleum ether, conc. phosphoric acid, glacial acetic acid, dioxane, phenol, acetone, ethylacetate, pyridine, acetophenone, formic acid	Insoluble	Insoluble

* The product was said to be insoluble if < 0.002 *g* of product were soluble in 100 ml of solvent at 24°. Extractions were performed by adding to 0.5 mg of authentic poly-L-phenylalanine and the ^{14}C-product of a reaction mixture (1800 counts/minute) 5·0 ml of solvent. The suspensions were vigorously shaken for 30 min at 24° and were centrifuged. The precipitates were plated and their radioactivity was determined.

In summary, a stable, cell-free system has been obtained from *E. coli* in which incorporation of amino acids into protein was dependent upon the addition of template RNA. Amino acid incorporation required ATP and an ATP-generating system, and was inhibited by puromycin, chloramphenicol and RNAase. Addition of poly-U resulted in the incorporation of phenylalanine alone into a protein resembling polyphenylalanine. Poly-U appears to function as a synthetic template, or messenger RNA, in this system. One or more uridylic acid residues appear to be the code for phenylalanine. Attempts are now being made to determine other letters of the code.

1961

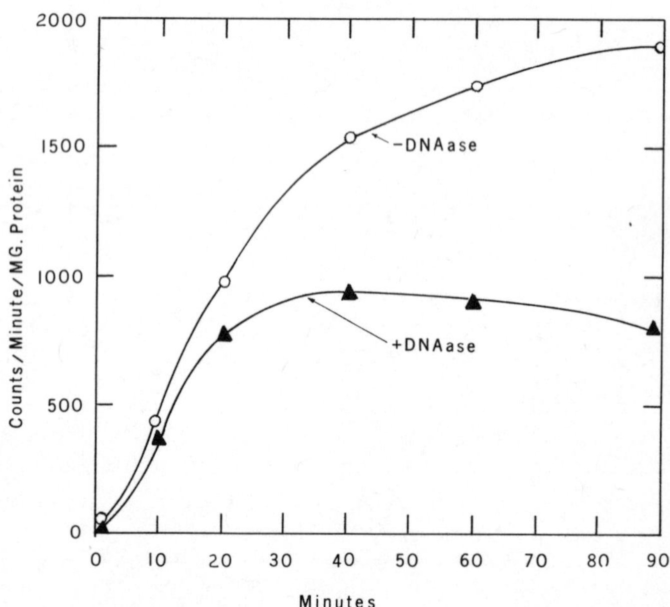

FIG. 1. The effect of DNAase upon ^{14}C-L-valine incorporation into protein.

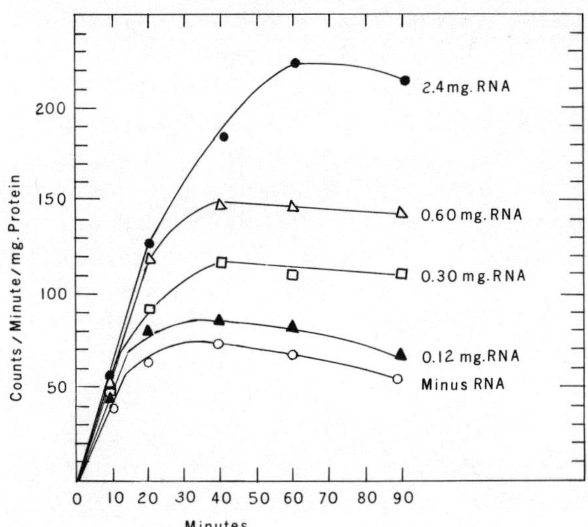

FIG. 2. Dependence of ^{14}C-L-valine incorporation into protein upon ribosomal RNA.

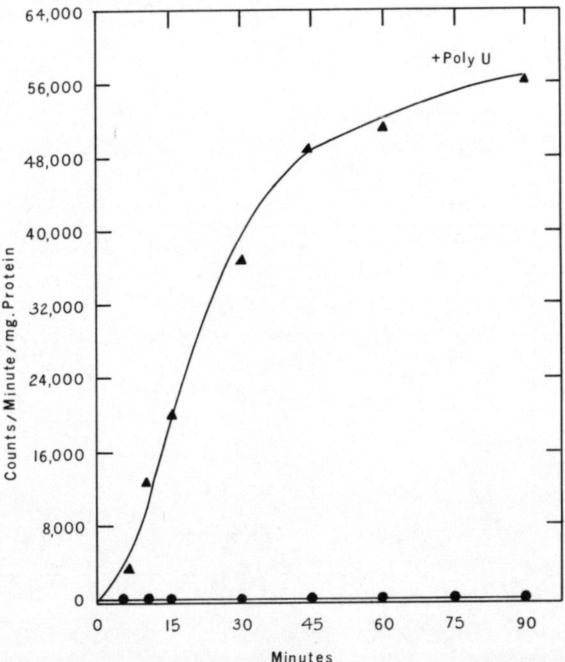

FIG. 3. Stimulation of U-^{14}C-L-phenylalanine incorporation by polyuridylic acid. ● Without polyuridylic acid; ▲ 10 μg polyuridylic acid.

PART V ○○○

Evolution

FROM ANTIQUITY TO THE present man has looked upon the origin of life in two distinct ways. One view represented in Western culture by *Genesis* looks upon all life as having a sudden origin in an act of creation by a Divine Being. The other looks on life as a process of change from simpler to more complex beings. Life, in this view, evolves. These two outlooks represent the two "purified" versions of special creation and evolution. Until the eighteenth century it was commonly believed even by the most religious of people that flies and worms arose spontaneously from decaying matter. This type of "spontaneous generation" was not considered contradictory to the theory of Special Creation, because these small organisms were usually too small to be seen and their complexity was not recognized. The definitive proof, by Louis Pasteur, that microbial life can only arise from preexisting microbes forced the scientist of the nineteenth century to choose either the purified version of special creation or that of evolution.

The first scientific attempt to construct a theory of evolution was Lamarck's theory of inheritance of acquired characteristics. Lamarck did not have any strong evidence for the evolution of life from simpler to higher forms. It was Charles Darwin who accomplished this. Darwin traveled through several regions of South America on a surveying voyage of the ship H.M.S. *Beagle*. As the official naturalist for the voyage, Darwin took extensive notes and made numerous collections of the flora and fauna he encountered. He was particularly impressed by the diversity of life on the Galapagos Islands, where organisms varied from island to island, and many forms occurred that were totally unknown on the west coast of South America. Darwin's observations gave him the impression that this diversity and relatedness of life must

have arisen by an evolution, but he had no theory to account for it. The needed theory came to him when he read an essay on human overpopulation by Thomas Malthus. Darwin reasoned that the animals he observed produced more progeny than could survive in their habitat. But which progeny survived? It was Darwin's belief that the most hardy, the "fittest," would survive. Hardiness, in turn was a feature of heredity. Organisms present a range of variations and the pressures of the environment permit the fittest to reach maturity. This phenomenon Darwin called *natural selection.*

Darwin's theory of natural selection was widely debated in the late nineteenth century, but by the beginning of the twentieth century most biologists accepted the general theory of evolution as part of their scientific outlook. Variations in populations were thought to form a continuous range, with imperceptible gradations distinguishing the members of a population. With the rediscovery of Mendelism, and Bateson's emphasis on discontinuous or sharply distinguished variation, there was some doubt about the *mechanism* of evolution. One of the major competitive theories in the early years of the twentieth century was Hugo De Vries's proposal that the origin of species was by sudden *mutations.* The "mutation theory" rejected Darwin's range of continuous variation, and it maintained that the new mutant population replaced the original population with a relatively uniform population; most of these new mutant individuals would be unfit, but the fittest forms would survive. De Vries used the primrose, *Oenothera,* to justify this theory because the original *Oenothera* was an import from North America and many new varieties had arisen and become established in Europe. The mutation theory was at first adopted by T. H. Morgan as his outlook on evolution. Using the fruit fly *Drosophila* to repeat De Vries's observations, Morgan found a different type of mutation. These "true" mutations caused changes of the various organ systems and characters of the fly, but they did not cause new species in one sudden mutational change. East, working with similar mutations in maize, found that the continuous variations of Darwin's observations could be accounted for by the activity of several genes. This "multiple-factor" or "polygene" theory was accepted by evolutionists as the basis of variation in Darwin's theory. By the late 1920's the work on the fruit fly had been sufficiently explored to justify a theory of *neo-Darwinism,* a view that tied the theory of

natural selection to the multiple-factor hypothesis and that attributed the origin of variation to gene mutations. The major spokesman for this view was H. J. Muller, who saw in the "individual gene" the basis for the origin of life and the basis for evolution.

In the 1930's a great many cytological studies were initiated to determine chromosome evolution in plants and animals. One of the most successful of these enterprises employed the giant salivary chromosome of the fruit fly. These chromosomes, about one thousand times larger than typical chromosomes, had a banded appearance which was characteristic for the species. Different species of *Drosophila* and different genera among the Diptera gave different salivary chromosome patterns. The closely related species still retained numerous sections of chromosomes which were homologous to one another. This was correlated, genetically, with the linear map order of the genes for these regions. The agreement between the sequences of the genes on chromosomes of different species strongly supported their common ancestry during their early evolution.

Evolution theory today no longer stirs debates among scientists as it did one hundred years ago. The tools of molecular biology are now providing a new study of evolution—the evolution of protein molecules. This approach has been extensively employed by Emil Smith and his colleagues for the cytochrome enzymes throughout cellular life from fungi to mammals. The work of Zuckerkandl and of Beuttner-Janusch has shown how useful the hemoglobin molecule is for an analysis of its evolution. In both instances, the replacements of amino acids involve mutational changes that probably occurred millions of years ago. No surprises have come from this work. The study of the evolution of proteins supplements the physiological and anatomical studies which have already provided a biological basis for classification and evolutionary relations.

Charles Darwin

Letter to Asa Gray, September 5, 1857

Darwin took an early interest in natural history. He collected insects; he enjoyed trips through the country; he was an ardent hunter. In his early twenties, he participated in an exploration of much of South America, including the Galapagos Islands off the west coast of Ecuador. These explorations and the discovery of the remarkable adaptedness of species to their environment led Darwin to a twenty-year scholarly analysis of his notes and of a vast quantity of technical publications by naturalists, breeders, farmers, geologists, and biologists. The idea that this adaptedness and distribution of animals could be interpreted as a consequence of evolution was always in his thoughts. But the mechanism eluded him and he refused to publish unsupported speculations. An essay by Thomas Malthus, an economist, on the overpopulation of the world, gave Darwin the mechanism he sought. Species produce more progeny than can be supported by their surroundings. Some must die. The survivors, Darwin asserted, must be more fit. A species was not composed of identical individuals but of a population with a range of differences. It was this variation which he had seen so frequently that provided the raw material for the environment to select. The survival of the best variants by *natural selection* became the mechanism of Darwin's theory of evolution. In 1859 Darwin published a book of some four hundred pages and entitled it *The Origin of Species.* This book was a much shortened version of a multivolume set which Darwin had planned for his theory. The letter to Asa Gray outlines the major points which he developed for this book.

After publication of *The Origin of Species* Darwin refrained from public debate. He lived a secluded life in his own estate at Down. Darwin was independently wealthy and he devoted his time to writing, correspondence, and numerous experiments with

Reprinted from *The Life and Letters of Charles Darwin,* ed. by Francis Darwin, London, 1888.

plants and animals. Eleven monographs and books followed the publication of *The Origin*. As his own critic, Darwin recognized that the weakest aspect of his theory of natural selection concerned variation. How did new variations arise? What accounted for the "continuous variation" characteristic of such traits as size and weight? Darwin hoped to find the answer in his own garden but his studies of pigeons, bees, and peas failed to reveal satisfying answers. It will always remain a source of historical amusement to speculate what would have happened if Darwin had read Mendel's paper on plant hybridization. A copy was sent by the editor to the Royal Society but Mendel was probably too humble a man to have requested that a copy be sent to Darwin. (For a recent view that Darwin would *not* have paid attention to Mendel's theories, see R. C. Olby's *Origins of Mendelism,* 1966, Constable & Co., London.)

Down, Sept. 5th

My dear Gray,—I forget the exact words which I used in my former letter, but I dare say I said that I thought you would utterly despise me when I told you what views I had arrived at, which I did because I thought I was bound as an honest man to do so. I should have been a strange mortal, seeing how much I owe to your quite extraordinary kindness, if in saying this I had meant to attribute the least bad feeling to you. Permit me to tell you that, before I had ever corresponded with you, Hooker had shown me several of your letters (not of a private nature), and these gave me the warmest feeling of respect to you; and I should indeed be ungrateful if your letters to me, and all I have heard of you, had not strongly enhanced this feeling. But I did not feel in the least sure that when you knew whither I was tending, that you might not think me so wild and foolish in my views (God knows, arrived at slowly enough, and I hope conscientiously), that you would think me worth no more notice or assistance. To give one example: the last time I saw my dear old friend Falconer, he attacked me most vigorously, but quite kindly, and told me, "You will do more harm than any ten Naturalists will do good. I can see that you have already *corrupted* and half-spoiled Hooker!!" Now when I see such strong feeling in my oldest friends, you need not wonder that I always expect my views to be received with contempt. But enough and too much of this.

I thank you most truly for the kind spirit of your last letter. I

agree to every word in it, and think I go as far as almost any one in seeing the grave difficulties against my doctrine. With respect to the extent to which I go, all the arguments in favour of my notions fall *rapidly* away, the greater the scope of forms considered. But in animals, embryology leads me to an enormous and frightful range. The facts which kept me longest scientifically orthodox are those of adaptation—the pollen-masses in asclepias—the mistletoe, with its pollen carried by insects, and seed by birds—the woodpecker, with its feet and tail, beak and tongue, to climb the tree and secure insects. To talk of climate or Lamarckian habit producing such adaptations to other organic beings is futile. This difficulty I believe I have surmounted. As you seem interested in the subject, and as it is an *immense* advantage to me to write to you and to hear, even so briefly, what you think, I will enclose (copied, so as to save you trouble in reading) the briefest abstract of my notions on the means by which Nature makes her species. Why I think that species have really changed, depends on general facts in the affinities, embryology, rudimentary organs, geological history, and geographical distribution of organic beings. In regard to my Abstract, you must take immensely on trust, each paragraph occupying one or two chapters in my book. You will, perhaps, think it paltry in me, when I ask you not to mention my doctrine; the reason is, if any one, like the author of the 'Vestiges,' were to hear of them, he might easily work them in, and then I should have to quote from a work perhaps despised by naturalists, and this would greatly injure any chance of my views being received by those alone whose opinions I value. . . .

I. It is wonderful what the principle of Selection by Man, that is the picking out of individuals with any desired quality, and breeding from them, and again picking out, can do. Even breeders have been astonished at their own results. They can act on differences inappreciable to an uneducated eye. Selection has been *methodically* followed in Europe for only the last half century. But it has occasionally, and even in some degree methodically, been followed in the most ancient times. There must have been also a kind of unconscious selection from the most ancient times, namely, in the preservation of the individual animals (without any thought of their offspring) most useful to each race of man in his particular circumstances. The "roguing," as nursery-men call the destroying of varieties, which depart from their type,

is a kind of selection. I am convinced that intentional and occasional selection has been the main agent in making our domestic races. But, however this may be, its great power of modification has been indisputedly shown in late times. Selection acts only by the accumulation of every slight or greater variations, caused by external conditions, or by the mere fact that in generation the child is absolutely similar to its parent. Man, by his power of accumulating variations, adapts living beings to his wants—he may be said to make the wool of one sheep good for carpets, and another for cloth, &c.

II. Now, suppose there was a being, who did not judge by mere external appearance, but could study the whole internal organisation—who never was capricious—who should go on selecting for one end during millions of generations, who will say what he might not effect! In nature we have some *slight* variations, occasionally in all parts: and I think it can be shown that a change in the conditions of existence is the main cause of the child not exactly resembling its parents; and in nature, geology shows us what changes have taken place, and are taking place. We have almost unlimited time: no one but a practical geologist can fully appreciate this: think of the Glacial period, during the whole of which the same species of shells at least have existed; there must have been, during this period, millions on millions of generations.

III. I think it can be shown that there is such an unerring power at work, or *Natural Selection* (the title of my book), which selects exclusively for the good of each organic being. The elder De Candolle, W. Herbert, and Lyell, have written strongly on the struggle of life; but even they have not written strongly enough. Reflect that every being (even the elephant) breeds at such a rate that, in a few years, or at most a few centuries or thousands of years, the surface of the earth would not hold the progeny of any one species. I have found it hard constantly to bear in mind that the increase of every single species is checked during some part of its life, or during some shortly recurrent generation. Only a few of those annually born can live to propagate their kind. What a trifling difference must often determine which shall survive and which perish!

IV. Now take the case of a country undergoing some change; this will tend to cause some of its inhabitants to vary slightly; not but what I believe most beings vary at all times enough for

selection to act on. Some of its inhabitants will be exterminated, and the remainder will be exposed to the mutual action of a different set of inhabitants, which I believe to be more important to the life of each being than mere climate. Considering the infinitely various ways beings have to obtain food by struggling with other beings, to escape danger at various times of life, to have their eggs or seeds disseminated, &c., &c., I cannot doubt that during millions of generations individuals of a species will be born with some slight variation profitable to some part of its economy; such will have a better chance of surviving, propagating this variation, which again will be slowly increased by the accumulative action of natural selection; and the variety thus formed will either coexist with, or more commonly will exterminate its parent form. An organic being like the woodpecker, or the mistletoe, may thus come to be adapted to a score of contingencies; natural selection, accumulating those slight variations in all parts of its structure which are in any way useful to it, during any part of its life.

V. Multiform difficulties will occur to every one on this theory. Most can, I think, be satisfactorily answered.—"Natura non facit saltum" answers some of the most obvious. The slowness of the change, and only a very few undergoing change at any one time answers others. The extreme imperfections of our geological records answers others.

VI. One other principle, which may be called the principle of divergence, plays, I believe, an important part in the origin of species. The same spot will support more life if occupied by very diverse forms: we see this in the many generic forms in a square yard of turf (I have counted twenty species belonging to eighteen genera), or in the plants and insects, on any little uniform islet, belonging to almost as many genera and families as to species. We can understand this with the higher animals, whose habits we best understand. We know that it has been experimentally shown that a plot of land will yield a greater weight, if cropped with several species of grasses, than with two or three species. Now every single organic being, by propagating rapidly, may be said to be striving its utmost to increase numbers. So it will be with the offspring of any species after it has broken into varieties, or sub-species, or true species. And it follows, I think, from the foregoing facts, that the varying offspring of each species will try (only a few will succeed) to seize on as many and as diverse

places in the economy of nature as possible. Each new variety or species when formed will generally take the place of, and so exterminate, its less well-fitted parent. This, I believe, to be the origin of the classification or arrangement of all organic beings at all times. These always *seem* to branch and sub-branch like a tree from a common trunk; the flourishing twigs destroying the less vigorous—the dead and lost branches rudely representing extinct genera and families.

This sketch is *most* imperfect; but in so short a space I cannot make it better. Your imagination must fill up many wide blanks. Without some reflection, it will appear all rubbish; perhaps it will appear so after reflection.

C. D.

P.S.—This little abstract touches only the accumulative power of natural selection, which I look at as by far the most important element in the production of new forms. The laws governing the incipient of primordial variation (unimportant except as the groundwork for selection to act on, in which respect it is all important), I shall discuss under several heads, but I can come, as you may well believe, only to very partial and imperfect conclusions.

1857

Hugo De Vries

The Principles of the Theory of Mutation

Hugo De Vries is one of the three co-rediscoverers of Mendel's laws. Before De Vries published this rediscovery in 1900 he had been an active theorist in heredity. In the 1890's De Vries proposed a theory of intracellular pangenesis in which he attributed specific traits to nuclear "pangens." In an attempt to solve the problem of variation, which Darwin had failed to do, De Vries sought an experimental system which would demonstrate changes or "mutations" in these pangens. His deliberate search for such a system was rewarded with the finding of abundant "mutations" in the evening primrose, *Oenothera lamarckiana*. Many of these sudden mutations were so pronounced and affected so many characters of the plant that De Vries thought that individual mutations could result in the formation of new species. In his theory of mutation all gradations of mutations are proposed, from trivial differences to profound changes requiring reclassification of the species. While looking for examples of mutation De Vries noticed that hybrids segregated certain traits. A detailed study of this led him to read Mendel's paper. It is not known if Mendel's mathematical treatment of these hybrid results independently occurred to De Vries or if he incorporated Mendel's interpretations into his own experiments and verified that they were true. Although De Vries received much fame for the rediscovery of Mendelism, he thought his greater fame was in the mutation theory. He accepted natural selection, but speciation he attributed to the sudden mutations rather than the gradual change brought by selection of continuous variations. The mutation theory was incorrect. In 1919 Muller proposed a mechanism for the origin of De Vries's "mutations" which did not involve the sudden changes he had advocated. Instead, they were a consequence of crossing over in complex chromosomal rearrangements. Large numbers of preexisting genes in these rearrangements were thereby made homozygous and could manifest their

Reprinted from *Science 40*:77–84, 1914, by permission of the publisher.

effects on several different tissues and structures. The evolution of *Oenothera* is a very special, unique case of chromosomal mechanisms which occasionally break down. Despite this unexpected outcome of the mutation theory, De Vries's contribution was significant. Among the newer generation of biologists were Morgan and his students, who studied mutation in *Drosophila* and found those "true" point mutations which do constitute the basis of evolution and Darwin's hypothesis of continuous variation.

Unity of internal structure combined with a great diversity of external forms is the great principle of organic differentiation. Lamarck was the first to point this out and to explain it by his theory of common descent. But the science of his time did not afford a sufficient body of facts in proof of his conception, and he failed to convince his contemporaries.[1]

It has been the work of Darwin to accumulate a large number of facts and arguments, borrowed from the most diverse parts of the physical and biological sciences, and to combine the main results of the study of nature in general in order to find a conclusive proof of the idea of Lamarck. Common descent is now acknowledged as the natural cause of the unity of organization. Successive slow modifications have produced the great diversity of forms and the diverging lines of evolution which have gradually led to the highest degrees of differentiation.

But his broad views and comprehensive considerations did not suffice to afford the desired proof. Comparative anatomy and systematical studies, the knowledge of the laws of the geographical distribution of animals and plants and of their gradual development during the geological epochs, could only outline the broad features of the theory. Evidently its basis must be sought in the study of the process by which one species is produced from another. Which is the nature and which are the causes of this process? Which are the elementary changes which, by numerous repetitions and combinations, have produced the main evolutionary lines of the animal and vegetable kingdom?

In order to answer these questions, Darwin studied the experience of the breeders. The improvement of domestic animals was well known at his time, the cultivated races of flowers and vegetables, of cereals and sugar-beets clearly and widely surpassed the same species in nature.

The method of breeders is based on the principles laid down about the middle of the last century (1840) by P. P. A. Leveque de Vilmorin, the father of the celebrated founder of the culture of sugar-beets. He had observed the high degree of variability of cultivated plants and discovered that by means of a choice of the best samples and by their isolation highly improved varieties may be produced. His son has applied this principle to the sugar-beets, one of the most variable of all cultivated forms, and succeeded in increasing the amount of sugar from 7 to 14 per cent. This improvement soon became the basis of a large sugar-industry in many countries of Europe. From that time isolation and selection have become the watchwords of a big new industry, which soon produced the most unexpected results in almost all parts of agricultural practice.

Darwin transplanted this principle of practice into pure science. He studied the variability of species in the wild condition and found it as widely spread and as rich in its features as in cultivated forms. He saw that very many species are distributed in nature in such a way as to constitute numbers of isolated colonies, sufficiently distant from one another to exclude the possibility of intercrossing. He discovered the great factor which replaces artificial selection in nature and called it by the name of natural selection. It is the unceasing struggle for existence and the victory of the most endowed individuals. In nature, every plant produces more seeds than can develop into new plants, owing to lack of space. Only those which are most fit for the surrounding conditions will survive, whilst the remainder are condemned to disappear. In this manner the struggle for life leads to a selection, which will be repeated in every generation, and a whole colony may gradually change by this means until at the end the characters are sufficiently different from the original ones to constitute a new variety or even an elementary species.

Natural selection in the struggle for life has now become the main principle of organic evolution. Since species obey in the wild condition the same laws as under cultivation, the principles of their improvement must be the same everywhere.

Darwin applied this principle to geological evolution also. Lyell had shown that the laws of nature have always been the same from the very beginning. Therefore natural selection must have been active from the first time of the existence of life on earth and have produced the main lines of differentiation as well

as the first traces of all those groups, which are now recognized as families and genera. It is my conviction that the success of Darwin in this line of ideas has been as complete as possible. He succeeded in convincing his contemporaries of the essential analogy between artificial and natural selection.

But, on the other hand, it must be conceded that the practice of breeders was not as simple as it seemed to be. No thorough study of the phenomena of variability had been made, and it was simply assumed that the diversity of forms within the cultivated races was due to one cause only. This was indicated by the well-known expression that no two individuals of a race are exactly alike. All specimens differ from one another in their industrial qualities as well as in their botanical characters. These qualities and characters are inheritable and the offspring of a selected individual will vary, according to Vilmorin, around an average lying between the type of the original species and that of the chosen individual. By this means the range of variability will be extended in the desired direction, and this may be repeated during a number of years, until the industrial value of the new race clearly surpasses that of the old one.

Evidently, it was said, natural selection must work in the same way. But the question remains whether this will really lead to new species, or only to local and temporary adaptations.

The answer to this question has been given by the newest discoveries of agricultural practice itself. Hjalmar Nilsson, the director of the celebrated experimental station of Svalöf in Sweden, discovered that variability among cultivated plants is not a single phenomenon, but contains at least two widely contrasted features.

He found that, apart from fluctuating variability, every cultivated species is a mixture of elementary types. A field of a cereal is only apparently uniform, and a closer investigation soon reveals numerous differences in the height of the stems, in the time of flowering, in the size and almost all other qualities of the ears, in resistance to diseases and especially in the industrial value of the grains. Moreover, he found that all these qualities are strictly inheritable. Nilsson took the grains of a single ear and found that all the individuals issuing from them are strictly alike and carefully repeat the characters of their mother. From such a chosen ear one may derive by repeated sowings grain enough to sow a whole field, and this will show an almost complete and

very striking uniformity. Therefore our ordinary species and varieties of cultivated plants are in reality mixtures of a smaller or larger number of different races, which grow together, but are, as a matter of fact, independent of one another. These races themselves are almost invariable, but their mixture in the field produces upon us the impression of a great variability.

What is the significance of this discovery for the explanation of artificial selection? Evidently this will tend to isolate the better races of the mixture and to exclude those of average or low value. Two methods may be followed. Either the breeder collects a handful of ears chosen with the utmost care from all parts of his field and secures a lot of grains large enough to sow a parcel of a moderate extent. Or he limits his choice to one ear only, which will take him one year more to obtain the necessary quantity of grains. The first method is the one which is still commonly followed, the second was introduced some twenty years ago by Nilsson.

The real nature of the first method may be explained by means of the careful studies of Rimpau, who applied it for the improvement of his rye. The group of ears of the first choice will evidently be itself a mixture, although of a lesser number of types. In choosing year after year a handful of the best ears Rimpau must gradually have purified this mixture, until after twenty years he succeeded completely in isolating the very best one of them. From this time his race must have been pure and constant, no further selection being possible. Using the method of Nilsson the same result may be reached by a single choice, and therefore in one year. The new race is produced by a jump and not by the slow and gradual improvement by small and almost invisible steps, which was assumed by Rimpau and Darwin.

From these discoveries the question arises, whether natural selection also proceeds by jumps and leaps, and not, as was commonly assumed, by imperceptible steps. The answer may be deduced from the observations of Jordan and others on the existence of elementary species in nature. Almost every wild species consists of some of them, and in special cases their number increases so as to embrace dozens or even hundreds of sharply distinguished types. Sometimes these are found in widely distant stations; at other times, however, they are growing in mixtures. Natural selection will, of course, under changed conditions, simply multiply one or two of the types to the exclusion of the others.

As a whole, the species will make progress in the desired direction, but in reality there will be no change of forms.

From all these and many other considerations it follows that the basis, which the practice of artificial selection seemed to afford to the theory of natural selection, is a fallacious one, and that the idea of evolution by means of slow and almost imperceptible steps must therefore be abandoned. But if this is conceded, how are species really produced in nature?

The theory of mutations answers that species are produced by means of jumps and leaps, exactly in the same way as varieties in horticulture. Varieties are only beginning species, says Darwin, and the same laws must govern the origin of both of them. Now, in horticulture, it is well known that varieties usually arise at once. In a field of a species with blue or red flowers some day an individual with white flowers is seen. Ordinarily it is only one, and it is not surrounded by transitions or by flowers of intermediate colors. Sometimes there may be two or three, but then their flowers are of the same degree of whiteness. One seed of the species has been transformed into a variety, and this is its whole origin. A single season suffices to produce the effect, no slow and gradual improvement being required. Moreover, the seeds of the first individual, if fertilized and saved separately, will reproduce the variety wholly pure. The same rule prevails for large groups of other cases; everywhere varieties arise by jumps, requiring only one year for their arrival.

The same rule also holds good in nature. But in order to show this, direct experiments are required. For this object I have cultivated a large number of wild species in my experiment garden, trying to see them produce varieties and to be enabled thereby to study the laws of this process. Let me adduce two instances, the origin of the peloriated toadflax and that of the double variety of the corn marigold. These varieties appeared in my cultures all of a sudden, after a number of years, the one in about half a dozen of individuals in successive generations, the other in a single instance. The ordinary toadflax has only one spur on its flowers and remained so in hundreds of individuals until a single specimen bore five spurs on every one of its flowers. The corn marigold had normal flower heads until 1899 when one individual produced some slight signs of duplication. Next year all its descendants bore double flowers and the race showed itself constant from the very beginning.

Thus, the production of varieties by leaps and jumps may be considered as a well-proved fact in horticulture and in a state of nature. It is a firm basis for a new theory, and we have only to transport the principle from the varieties to the origin of elementary species. Recognized for species, the theory will obviously be true for genera and families also, and explains the evolution of all organic beings in all the different lines of the genealogical tree.

The idea of the origin of species by leaps and jumps has the great advantage of answering in an unexpected and decisive way the numerous and in part very grave objections which have been brought forward against the theory of Darwin. To my mind, this is one of the best arguments in its favor. It releases the theory of evolution from the serious difficulties which its adversaries have never ceased to urge against it. Therefore it seems useful to give a brief survey of them now.

The oldest and most serious objection is based on the obvious uselessness of new characters during the first stages of their evolution, if this is supposed to be invisibly slow. Imperceptible odors can not guide insects in their visits to flowers and assure to these a sufficient advantage in the struggle for life. Adaptations for the capturing of insects by plants would be of no value in a primary and imperfect condition and therefore can not be evolved by the action of natural selection. Imperfect instincts would be rather obnoxious, according to Wasmann, and thus would be liable to be destroyed instead of increased by this action. So it is in many other cases. Beginning characters would always be too insignificant to be of any value in the struggle for life. Evidently the principle of leaps and jumps at once relieves us of the necessity of this hypothesis. It does not admit a gradual appearance of characters, but assumes these to appear at once in the full display of their development, and without the aid of natural selection.

The same holds good for useless characters. The theory of Darwin can not explain them. According to him, every quality is developed exactly through its utility, and useless properties should be eliminated from the very beginning by the struggle for life. But it is now generally recognized that many beautiful differentiations are in reality no adaptations at all, and that their usefulness is at least very doubtful. This, for instance, is the case of heterostyly and of the likeness of the flowers of some

orchids to insects. The theory of mutations has no difficulty with useless and even with slightly prejudicial characters. Arising by a sudden jump, they may keep their place, provided only that they are not in such a degree hurtful as to prevent a normal development of the individuals.

A third objection has been derived from the studies of the celebrated anthropologist Quételet, who discovered the general law of fluctuating variability. He introduced the principle of studying every quality for itself and of comparing the different degrees of its development in a large number of individuals. He found by this means that characters simply follow the laws of probability. They vary around an average condition in two directions, of increase and decrease, but precisely thereby this variability excludes the production of a new character. Darwin tried to derive the one from the other, whilst the theory of mutations recognizes the almost diametrically opposed nature of the two phenomena.

A last objection has been brought forward by the study of the age of the earth. Physicists as well as astronomers have refused to accept the theory of slow evolution as the time required by Darwin in connection with his ideas, seemed by far too long. A man's life would not suffice to see the changes, which, after him, would be necessary to produce a single step in the line of evolution. The differentiation of a flower or of a seed would require millions of years if it went on so slowly, and the development of the whole organism of a plant, and still more so that of the higher animals, would obviously require a vastly larger amount of time. Darwin has calculated the necessary time for the evolution of the whole animal and plant kingdom on the assumption of slow and almost imperceptible changes, and estimated it to be at least equal to some thousands of millions of years.

But our globe can not be as old as that. There is quite a large number of arguments which allow us to estimate the age of the earth with a sufficient degree of accuracy, and they all point, unanimously, to a period of only some twenty or forty millions of years. This number is evidently far too small for the explanation given by Darwin and in consequence thereof it has always been considered as one of the most decisive arguments against the theory of slow and gradual evolution.

In order to estimate the age of the earth different phenomena may be used. First the separation of the moon, secondly the

solidification of the earth's crust, then the condensation of the aqueous vapor and the formation of oceans. The quantity of salt dissolved in these oceans and the thickness of the geological layers, especially those of a calcareous nature, afford further arguments.

According to George Darwin the moon was separated from our globe about 56 millions of years ago. The age of the solid crust has been calculated by Lord Kelvin from the increase of the temperature in deep mines. In some regions the temperature is seen to increase about one degree for every fifty meters; in others, however, one degree for a hundred meters. On the average the considerations of Lord Kelvin gave an age of twenty to forty millions of years for the solid crust of the earth.

The quantity of salt obviously increases in the oceans on account of the salt added by the rivers and of the evaporation of the water. The total quantity of this salt has been calculated and the quantities of the yearly supply of water are known for all the larger streams, as well as their percentage of salt. From these data we may calculate the annual increase of salt in the oceans and find how many years would be required for our present rivers to accumulate all the salt now found in the seas. According to Joly, about ninety millions of years would be necessary. But obviously the rivers must exhaust the grounds which they drain, and formerly these must therefore have been much richer in salts. This consideration must lead us to diminish the number of years required in a very sensible manner.

The age of the geological strata has been deduced from their thickness and the velocity of the process of sedimentation. Sollas estimates the total thickness at about 80 kilometers and the average rate of deposition of the layers at 30 cm. per century. From these numbers we may find an age of 26 millions of years for the collective deposition of all the geological layers. Calcareous rocks have been built by organisms and mainly by corals and molluscs. These have made use of the lime added to the sea by the rivers. Dubois has calculated on the one hand the whole thickness of these rocks and on the other the yearly supply of lime from the rivers. He concludes that 36–45 millions of years would be required to produce the whole of this system.

All these data have been subjected to a criticism by Sollas and compared with one another. Obviously the highest estimates are only limits, and in considering this, Sollas arrives at the general

average of about 20–40 millions of years. He points out that the epochs which have served as starting points are not very far distant from one another, considered in a geological way, and that therefore they may be taken together to delineate the duration of organic life on this earth.

As we have seen, this duration is by far too short to allow the slow and gradual development of life supposed by Darwin. It necessitates a very substantial abbreviation of this process and thus affords one of the best supports of the theory of mutations.

Thus we see that this theory is based on almost all the branches of natural science. All of them join in the assertion that the hypothesis of slow and almost invisible changes is too improbable to be accepted and is even in open contradiction to some of the best results of other sciences. The theory of an evolution by leaps and jumps evades all these objections and thereby releases the theory of Darwin from its separate position.

But it is doing more than this. By rejecting the hypothesis of invisible changes it leads us to search for the visible alterations, which it assumes to be the leaps and jumps by which animal and vegetable species are being produced. If the transformation of one species into another is a visible process, it must evidently be sought for and be brought to light in order to study its laws, and to derive from this study an experimental proof for the theory of evolution.

However, it is hardly probable that these jumps are numerous in nature as it now surrounds us. On the contrary, they must rather be rare, since nobody has seen them until now in the field. Therefore I have sought for a plant which would produce more of such mutations than other plants. I have studied over a hundred species, investigating their progeny, and among them one has answered my hopes. This is the evening primrose of Lamarck, which chances to bear the name of the founder of the theory of evolution which it is prepared to support. It is a species which grew wild in the territory of the United States, where it has been collected by the well-known traveler and botanist Michaux, and whence Lamarck derived the authentic specimen for his description. Since that time it has spread in Europe and is now found especially in England, Belgium and Holland in a number of localities, some of which consist of many thousands of individuals. In more than one of these localities it has been ob-

served to produce mutations, especially in a field near Hilversum in Holland, whence I have obtained the individuals and seeds which have served as the starting points of my cultures.

In these cultures the species is seen to be very pure and uniform in the large majority of its offspring, but to produce on an average one or two aberrant forms in every hundred of its seedlings. The differences are easily seen even in young plants and are mostly large enough to constitute new races. The more common ones of these races are produced repeatedly, from the seed from the wild plants as well as in the pure lines of my cultures. It is obviously a constant and inheritable condition which is the cause of these numerous and repeated jumps.

These jumps at once constitute constant and ordinarily uniform races, which differ from the original type either by regressive characters or in a progressive way. By means of isolation and artificial fecundation these races are easily kept pure during their succeeding generations.

I shall not insist here upon their special characters. The most frequent form is that of the dwarfs, *Oenothera nanella,* and the rarest is the giant, or *O. gigas,* which has a double number of chromosomes in its nuclei (28 instead of 14) and by this mark and its behavior in crossing proves to be a progressive mutation. Other new types which are produced yearly are *O. rubrinervis, O. oblonga* and *O. albida. O. lata* is a female form, producing only sterile pollen in its anthers and *O. scintillans* is in a splitting condition, returning every year in a greater or less number of individuals to the original type from which it started. Besides these there are a large number of mutations of minor importance, many of which have not even been described up to the present time.

Thus we see that the experiments provide us with a direct proof for the theory of evolution. They constitute an essential support of the views of Darwin, and moreover they relieve them of the many objections we have quoted and bring them into harmony with the results of the other natural sciences.

But, besides this, they show us the way into a vast new domain of investigation and afford the material for a study of the internal and external causes which determine the production of new species, at least in those cases in which, as in the primroses, mutations are relatively abundant. From these we may confidently

hope to come some day to the study of those rarer mutations on which the differentiation of the main lines of organic evolution seem to have depended.

1914

NOTE

1 "The Mutation Myth" is the title of a recent article in this journal, N.S., Vol. XXXIX, No. 1005, April 3, 1914, p. 488. Its author, Edward C. Jeffrey, starts from the conception that the mutation theory has been derived from my experiments with *Oenothera Lamarckiana* and allied species. This opinion is indeed, even yet, not unfrequently held by those who have not read my books. It is obviously erroneous and therefore may well be called a myth. Logically and historically the desirability of those experiments has been derived from the theory, as will be seen in the text. Jeffrey bases his arguments upon the well-known researches of Geerts concerning the partial sterility of many of the members of the natural family of the *Onagraceae*. Geerts found that in almost all the genera of this family, including all their species as far as investigated, the ovules are for one half in a rudimentary condition, which excludes the possibility of their being fertilized, whilst about one half of the pollen grains is sterile. This double character has therefore persisted during the pedigree-evolution of almost this whole family. In contradiction with Geerts, Jeffrey considers it to be an indication of a hybrid condition. If this were true, almost the whole natural family of the *Onagraceae* would have evolved in a hybrid condition and *Oenothera Lamarckiana* would follow the rule. It remains doubtful, however, how this hypothesis could explain the high degree of mutability of *O. Lamarckiana,* since the majority of the supposed hybrid species do not show signs of such a condition.

T. H. Morgan

For Darwin

Morgan's career as a biologist covered many fields. He performed experiments with more than fifty different organisms. Embryology was his major interest but at the turn of the century he, like all biologists, was forced to evaluate the newly discovered laws of heredity. Mendelism seemed incompatible with Darwinism and Morgan saw in De Vries's mutation theory a possible solution for the theory of evolution. This essay illustrates the skepticism characteristic of Morgan's outlook; as his student, H. J. Muller once remarked, Morgan "doubted the doubt until he doubted it out." Morgan could not accept the Lamarckian interpretation of acquired characteristics. He was, nevertheless, an avowed mechanist—there was no direction or purpose imposed on the mutation process. Rather, Morgan accepted Darwin's view of natural selection as the basis for the survival of the most adaptable mutations. It was probably this deep appreciation of De Vries's work which led Morgan to search for mutations in the fruit fly *Drosophila* shortly after this paper was written. Morgan also attempted, unsuccessfully, to induce mutations with radium, but he was not convinced he had demonstrated any new mutations. Ironically the mutations leading to new species or varieties which De Vries found in *Oenothera* were not found in *Drosophila.* Indeed, the mutations found by De Vries were not mutations as we understand that term today. De Vries's mutations were complex chromosomal changes brought about by crossing over with aberrant chromosomal arrangements. But the mutations found by Morgan and his students in *Drosophila* did lead to a neo-Darwinian theory of evolution which is the most widely accepted interpretation today.

The following was a lecture on "Darwin's Influence on Zoology," delivered at Columbia University, February 26, 1909.

Reprinted from *The Popular Science Monthly,* April, 1909, pp. 367–380.

We have come together to-day to consider Darwin's influence on zoology. It is a hazardous task to pretend to estimate the influence of any event on the course of history so long as we can not know what the outcome had been otherwise. But to this at least we can testify, that it is the general belief of zoologists to-day that Darwin's influence in bringing about the acceptance of the theory of evolution marks a turning point in the history of their science, and I shall attempt to justify this opinion by pointing out the condition of zoology before Darwin and its subsequent course of development after 1859.

To the zoologist Darwin was above all else a zoologist. It is true he interested himself greatly in geology, but he does not stand as a leader of that science; he carried out many experiments with plants and wrote some important botanical books, and here the zoologist will yield second place to his brother, the botanist. Darwin wrote on the "Descent of Man," he studied the expression of the emotions and carried out physiological work along several lines, yet I should not rank him preeminently an anthropologist, a psychologist or a physiologist any more than a paleontologist or a botanist.

In the mind of the general public Darwinism stands to-day for evolution. The establishment of the theory of evolution is generally accepted as Darwin's chief contribution to human thought, and while Darwin did not originate this idea that forms the framework of our modern thinking, yet by general accord its acceptance is attributable, and justly so, to Darwin.

To the zoologist Darwinism means more especially evolution accounted for by the theory of natural selection, yet also many other things, to which I shall refer in the proper place.

But I shall attempt this afternoon, before all else, to convince you that the loyalty that every man of science feels towards Darwin is something greater than any special theory. I shall call it the spirit of Darwinism, the point of view, the method, the procedure, of Darwin.

In order that we may form some idea of Darwin's influence on zoology, let us examine the condition of that science prior to 1859 to see what influence zoology had on Darwin and his contemporaries. I shall not try your patience by attempting to review the history of the subject, but it would not belittle the greatness of Darwin's achievement one whit to find that brilliant discoveries had been made before his time, the theory of evolution

plainly enunciated, the doctrine of spontaneous generation disproved; comparative anatomy widely studied; the important functions of the body elucidated, the foundations of the science of embryology laid, and the principles of pedigree breeding followed.

In the eighteenth century, when the study of different kinds of animals inhabiting sea and land attracted the attention of zoologists, great classifications were invented. Two main facts emerged. On the assumption of fixity of type, a classification of the different forms of animals and plants became possible. But on the other hand the more extensive the material to be classified, the more difficult it became to make such systems, for the fixity of type was often lost in apparent transitions to other types. Counter claims arose as to the superiority of one system over another, and the question of an artificial system versus a natural one was widely debated. Now, an artificial system, like the arrangement of the words in a dictionary, is obviously only a matter of convenience, but it became a question of deep philosophical importance to decide what was meant by a natural classification. To us at the present time a natural classification implies a relation due to descent; it is neither more nor less than the natural relation of a man to his ancestors. But it was a fatal mistake to read our meaning backwards to the time before Darwin.

To the great Cuvier a natural system meant an assemblage of groups having a common plan of structure, and he was enraged by Geoffroy St. Hilaire's attempt to put all animals from the bottom to the top in a straight line. A common plan of structure might only mean that idea which best expressed the outcome of a wide study of structure; but to those who tried to peer behind the scenes it meant not seldom to fathom the creation of the world; and it required no vivid imagination to add that it gives an insight into the plan by which the world was created.

A historian of the times wrote:

"Yet in fact the assumption of an end or purpose in the structure of organized beings appears to be an intellectual habit, which no efforts can cast off. It has prevailed from the earliest to the latest ages of zoological research, appears to be fastened upon us alike by our ignorance and our knowledge . . . and the doctrine of unity of plan of all animals, and the other principles associated with this doctrine, so far as they exclude the conviction of

an intelligible scheme and a discernible end, in the organization of animals, appear to be utterly erroneous."

Contrast, in passing, this pious conviction with Geoffroy's modest lines:

"I ascribe no intention to God, for I mistrust the feeble powers of my reason. I observe facts merely and go on. I can not make nature an intelligent being who does nothing in vain, who acts by the shortest mode, who does all for the best."

Thus arose in the eighteenth and nineteenth centuries the dogma of the fixity of species—a dogma based, it is true, on a direct appeal to fact as well as to conscience. But this dogma contained the germ of its own undoing, in so far as it appealed for its support to observations that every man might make for himself. Yet so influential were its advocates, so convinced of its truth, that more than one assault was made before it crumbled away.

It is no small pleasure to repeat to-day the names of those bold and original thinkers, who braved the displeasure of their compatriots and the contempt of their times, who brought forward evidence and argument to disprove the teaching of the schools. Their work, it is true, failed in the sense that it received no sufficient meed of praise or word of commendation, but who will deny that a seed was sown that in time bore fruit? Foremost, I think, ranks the great Lamarck, the centenary of whose "Philosophie Zoologique" is celebrated this year in France—a bold spirit, whose ideas, based on a wide familiarity with facts, live and bear fruit to-day. Geoffroy St. Hilaire, advanced thinker and philosopher of nature, opponent of the great anatomist Cuvier, brought the problem of evolution to the bar of judgment, losing the decision, it is true, but his ideas a later generation hold in high esteem. Erasmus Darwin, grandfather of our Darwin, author of "The Zoonomia," celebrated in verse "The Botanic Garden" and the "Loves of the Plants," and even before Lamarck, advocated the principle of evolution and the theory of inheritance of acquired characters. Herbert Spencer, adopting the idea of evolution, laid thereon the elaborate superstructure of his philosophy. Robert Chambers, too, kept alive the central idea of change in the organic world in his "Vestiges of Creation." Others there were, besides, in different lands, but these especially were nearer to Darwin and his times.

We come now to the years between 1837 and 1844, when Darwin

was making his memorable notes on the relation between varieties and species. Reading through his letters of this period one is surprised to find how little he was impressed by the history of zoology and the influences of his own time, and how much he based his conclusions on the results of his own close observations, his accumulation of data, and careful consideration of facts. In regard to Lamarck, Darwin states in his autobiography, that in 1825 when he was at Edinburgh University, Dr. Grant "burst forth in high admiration of Lamarck and his views on evolution. I listened in silent astonishment, and as far as I can judge, without any effect on my mind."

In later years, after reading Lamarck, Darwin wrote Lyell, in 1859:

"You often allude to Lamarck's work; I do not know what you think about it, but it appeared to me extremely poor; I got not one fact or idea from it."

Writing to Lyell in 1863, he says:

"You refer repeatedly to my view as a modification of Lamarck's doctrine of development and progress. . . . Plato, Buffon, my grandfather before Lamarck, and others, propounded the *obvious* views that if species were not created separately they must have come from other species, and I can see nothing else in common between the 'Origin' (of Species) and Lamarck."

Darwin wrote to Hooker in 1844:

"Heaven forfend me from Lamarck's nonsense of a 'tendency to progression,' 'adaptations from the slow willing of animals,' etc. But the conclusions I am led to are not widely different from his; though the means of change are wholly so."

Darwin had read "The Zoonomia" of his grandfather prior to 1825 in which "similar views (to those of Lamarck) are mentioned but without producing any effect" on him. He continues, with his usual candor:

"Nevertheless it is probable that the hearing rather early in life such views maintained and praised may have favored my upholding them under a different form in my 'Origin of Species.' "

It is a regrettable fact that Darwin did not appreciate Lamarck's work. The failure of Lamarck's writings to produce any apparent influence on Darwin may be attributed, I think, to the form in which Lamarck's views are presented. He uses facts as illustrations of his ideas, while with Darwin the facts are all important as furnishing the evidence on which a theory is to be

established. He misunderstood Lamarck's view in regard to the inheritance of acquired characters, yet held himself the same opinion in the main as had Lamack. The modern idea of descent, as a system of branching due to *divergence* in those species descended from the same parent species, was expounded luminously by Lamarck, yet Darwin discovered it independently for himself. He says:

"But at that time (1844) I overlooked one problem of great importance; and it is astonishing to me . . . how I could have overlooked it and its solution. This problem is the tendency in organic beings descended from the same stock to diverge greatly in character as they become modified. That they have diverged greatly is obvious from the manner in which species of all kinds can be classed under genera, genera under families, families under suborders, and so forth, and I can remember the very spot in the road where to my joy the solution occurred to me."

It is the same view that Lamarck had fully expounded thirty-five years before.

We have now arrived at the period just before the publication of Darwin's famous book. It is sometimes said that the time was ripe for the reception of the ideas formulated by Darwin—it was in the air, as we say—but if so, it must have been so attenuated as to be invisible to eyes as sharp as Huxley's and the other famous naturalists of that time. Huxley says that within the ranks of the biologist he met with but one who had a word to say for evolution. Outside these ranks the only person known to him "whose knowledge and capacity compelled respect" and who advocated evolution was Herbert Spencer. "Many and prolonged were the battles they fought" on this topic, but Huxley maintained his agnostic position. He states:

"I took my stand upon two grounds; firstly, that up to that time the evidence in favor of transmutation was wholly insufficient; and secondly, that no suggestion respecting the causes of the transmutation assumed which had been made was in any way adequate to explain the phenomena. Looking back at the state of knowledge at that time I really do not see that any other conclusion was justified."

This frank statement of Huxley not only gives us an insight into the position of one of the most progressive zoologists of that time, but what is of more importance it implies also *why* the "Origin of Species" convinced him of the doctrine of evolution.

We have now sufficiently traced the possible influences of the times on Darwin. Before we proceed to study the influence of Darwin on his time, let us for a moment pause to consider what influence Darwin's own surroundings had in shaping his views. His voyage in the *Beagle* had brought him in contact with the question of geographical distribution. He read Malthus in 1838 and this gave him his first idea of the survival of the fittest; and, as his son and biographer states, this date marks "the turning point in the formation of his theory," so that by 1844 he formulated "a surprisingly complete presentation of the argument afterwards familiar to us in the 'Origin of Species.' " His extensive study of variation under domestication furnished him with the experimental evidence that went so far towards making his study of variation of far-reaching and profound importance. Indeed, in this one essential respect, Darwin was far ahead of all of his contemporaries, and, if you will pardon the anachronism, far ahead of his successors. It is only in recent years that zoologists and botanists have begun once more to work the rich mine of materials at their very doors. The paper of Wallace on "Natural Selection" in 1858, the reception to the "Origin of Species" in 1859, the storm of disapproval it met on the one hand, the staunch and able friends it made on the other, need only be recalled in passing.

We come now to the influence that Darwin's work has had on modern zoology. That influence is due not alone to the "Origin of Species" that gave to the world an abstract only of his views, but equally to his other works, especially, I think, the "Variation of Animals and Plants under Domestication," and the "Descent of Man." After Darwin and largely as an outgrowth of the wide interest his views aroused in all branches of zoology we find activity going on in many lines of work. One group of workers, the systematists, have kept nearer, I think, to the older traditions. They have been concerned with three of the most important matters that have a direct influence on the "Origin of Species"—the intensive study of species and varieties, the geographical and geological distribution of animals, and the influence of the environment in modifying species. Their results have supplied the most extensive contributions, perhaps, that have been made to the theory of species-formation and transmutation. They seem to me, however, to have paid less attention to another, equally important, field, that of the adaptation of animals to their environ-

ment, and the causes that have been effective in bringing about this adaptation. To physiology we look in vain for an answer to this question, that is perhaps a physiological problem, for while physiology has advanced to a wonderful degree our knowledge of the complicated adjustment within the body, the origin in time of these adjustments and their relation to the outer world has excited less interest.

The morphologists, or philosophical anatomists, form the second great group of students whose activity is a direct outgrowth of Darwinism. The determination of the relationships of the great classes of animals on the principle of descent has occupied much of their time. Two other important fields of labor have also fallen to their share. The study of development or embryology has been almost exclusively pursued by morphologists, inspired in large part by the theory of recapitulation.

The older form of the doctrine, that in the development of the individual the past history of the race is repeated, has been revived —a doctrine much in vogue in the early part of the last century, which has continued to have its followers despite the different interpretation that von Baer gave to the same facts. Whatever interpretation we choose at the present time, the presence of structures like gill-slits in the human embryo, directly comparable to those in the fish, has had an important influence in disentangling the relationship of living animals to their remote ancestors.

The morphologist has also undertaken the study of heredity, and the relation of heredity to the germ-cells that are the links in the chain of organic life. Few other studies have advanced in recent years at a more rapid pace and few have yielded facts of greater significance, for here lies the key to the origin and nature of variations.

Systematists and morphologists alike have been evolutionists, but it is a curious fact of zoological history that until very recently there has been no body of students whose interests have been directed *primarily* towards the problems of evolution. This is due, I think, to a general feeling that the data for evolution are rather the by-products of the zoologist's work-shop, than products directly manufactured by him, despite the splendid example of Darwin to the contrary. Is it not strange, therefore, with all the real interest in the theory of evolution, that so few of the immediate followers of Darwin devoted themselves exclusively to a study of that process? As I have said, the systematists have been

accumulating a vast amount of valuable material, but their chief interest has, on the whole, been in its classification, only secondarily in its bearing on evolution. The morphologist has been busy in *applying* the theory of evolution to the explanation of group relationships. The paleontologist has perhaps been more directly concerned with the evolution question than has any other worker.

There is a school that calls itself Lamarckian or Neo-Lamarckian which as far as its name goes should include the followers of Lamarck rather than of Darwin. Yet with few exceptions the Neo-Lamarckians derive their inspiration, I think, directly from Darwin. Darwin held that characters acquired during the lifetime of the individual may be transmitted to the offspring. He abhorred what he supposed to be Lamarck's rubbish, that an animal acquired a new part by willing it. We have seen that this is a travesty on Lamarck's real teaching, and that on the whole Darwin's view of acquired characters is almost Lamarck's. Yet the modern Lamarckians get their doctrine directly from Darwin rather than from Lamarck, who propounded it fifty years earlier, as had Erasmus Darwin, still earlier.

I have laid emphasis on the relation of Lamarckism to Darwinism in order to draw attention to the problem of adaptation. The Neo-Lamarckians have kept this all-important question in the foreground, while others have taken adaptation too much for granted in their attempts to explain the origin of species; for species, in a technical sense, may have little to do with the problem of adaptation. The life of an animal is intimately dependent on its adaptive characters, but its "specific characters" may be largely unimportant for its existence.

Systematists and morphologists include broadly the followers of Darwin during the thirty years after the publication of the "Origin of Species." They have advanced to a high degree the principles of their science, and the modern aspect of zoology is largely the outcome of their varied and far-reaching labors.

There is a small group of writers scattered amongst these larger groups that are ranked or rank themselves Neo-Darwinians. I must pause a moment to pay them my tardy respects. They have set themselves up to be the true Darwinians. They seem less concerned with the advancement of the study of evolution than with expounding Darwinism as dogma. Their credulity is more remarkable than their judgment. To *imagine* a use for an organ is for them equivalent to *explaining* its origin by natural selection

without further inquiry. Any examination, in fact, into the nature of variation, they appear to regard as superfluous, although harmless, but it is heresy to study critically the working out of the theory of natural selection. Such has ever been the procedure of infertile followers of great leaders. In the present instance the result is the more deplorable, since Darwin's own independence of the traditions of all schools, his careful study of facts, his emancipation from prejudice, are his lasting virtues. The Neo-Darwinian, worshipping the letter of the law, forgets its import. Let us salute, and pass.

And now we come to the last twenty years of zoology as influenced by Darwin. This, I believe, is the brightest chapter of Darwinism, for the spirit of Darwin is once more abroad.

Foremost amongst the many debts that modern zoology owes to Darwin is this: he pointed out that in order to understand how evolution takes place, we must study the variations of animals and plants, for here is the material on which rests any solid superstructure. To my mind, the appreciation of this maxim and its application is the distinguishing feature of Darwin's work. Before his time the theory of evolution remained but a general idea, though one of profound significance. After Darwin, the theory of evolution rested its claims for recognition on a definite body of information relating to variations and their inheritance. It is these data that first convinced his greatest contemporaries of the reality of evolution, and finally convinced also the rank and file of thinking men. So extensive were the facts of variation accumulated by Darwin, so penetrating was his analysis of these facts, so keen was his insight, and so wise his judgment as to their meaning, that for thirty years afterwards little of importance in this direction was added. In their amazement at Darwin's accomplishment zoologists forgot that he had opened the door leading into an unexplored territory. During the last twenty years the march forward has once more begun and the reward has been immediate.

Let us tarry therefore a little in these rich and pleasant fields of discovery and examine in some detail what is being done. The study of variation has been actively pursued in three main directions. The biometricians have applied exact measurements to variation; the ecologists have studied the complex influences of the environment; the experimentalist has put to the test the

supposed factors of change. Each of these methods has brought out results of significance.

A careful study of variations within each species has shown that taken as a group many variations conform to the law of probability. Popularly expressed, this means that chance determines variations, or, put more exactly, variations taken as a group and measured, give the same mathematical results that follow when any set of objects become arranged according to the laws of probability. There was a time when chance meant lack of conformity to law. Such a popular interpretation has no scientific standing. The great law of causation is not abrogated, but the outcome is only the result of a large number of small influences whose effects depend on the nature of the material and on the nature of the conditions. It is so important that this fact be clearly understood that I may be pardoned if I call to mind some familiar illustrations. No two leaves on a tree are identical, yet if many are measured, they give the curve of probability. Men are of different heights, yet they range about a mode. Color appears in various shades, yet if standardized, it is found to follow the same laws of chance variation.

What value have these facts for the theory of evolution? If in every generation we find that the same kinds of individuals recur, the results mean stability, not progress. That this state of affairs actually exists in many species living under the same environment during successive generations there can be little doubt. But change the environment and the results also change. Another factor comes to light that is independent of outside conditions. It is what has been called preferential mating. If within a group the males and females of certain kinds tend more often to pair with each other, the collective group becomes modified in one or more directions. In man this factor assumes a special importance, for, as Pearson has shown, there is measurable evidence that such mating occurs.

It has often been urged, and I think with much justification, that the selection of individual, or fluctuating variations could never produce anything new, since they never transgress the limits of their species, even after the most rigorous selection—at least the best evidence that we have at present *seems* to point in this direction. But a new situation has arisen. There are variations within the limits of Linnean species that are definite and inherited, and there is more than a suspicion that by their presence the possibility is assured of further definite variation in the same

direction which may further and further transcend the limits of the first steps. If this point can be established beyond dispute, we shall have met one of the most serious criticisms of the theory of natural selection.

It is not without interest to note in this connection that Darwin often assumed that *fluctuating variations* are transmitted to the offspring. The idea that they are not was a later development—the result, it is true, of a better knowledge of the law of fortuitous effects, or of probability. But we have discovered the additional fact that *some small* variations are inherited. Let us call these *definite variations,* and if these be the material with which evolution is concerned, Darwin's assumption in regard to the nature of variation will be, in part, justified.

These small, definite variations appear to be closely allied to those larger, more visible definite variations that we now call mutations. We owe our modern ideas of such variations mainly to De Vries and to those who have followed in his footsteps. Such sudden changes have been long known and were spoken of by Darwin as saltations—or sports. Darwin knew of cases like the ancon ram, from which a race of short-legged sheep was produced. He knew that totally black or melanistic mutations and albinos arise in many groups suddenly, and transmit their characters. A black-shouldered or japanned peacock has appeared more than once and perpetuated itself without selection. It would be out of place to-day to discuss this absorbing problem. That extreme mutations may at times have been an element of progress in nature few will deny, especially if we exclude such monstrous forms as those the breeder has used in building up domesticated races of animals.

It is not, however, to these extreme examples of definite variations that I wish especially to draw your attention, but to that group of smaller variations of a similar nature that may at their first appearance fall within the limits of ordinary variability. I now ask you, therefore, to follow me in an attempt to apply this latest discovery to the theory of evolution.

If we trace the ancestors of any living animal—man, for example—we discover that his ancestry goes back not as a single line, nor as a converging system of lines, but as a vast branching network. Each man has had 2 parents, 4 grandparents, 8 great-grandparents, 16 in the fourth generation, 32 in the fifth, 64 in the sixth, 128 in the seventh, 256 in the eighth, 512 in the ninth,

1,024 in the tenth. A few generations further removed we should expect to find that the majority of all the individuals of the species had poured their blood, as we say, into each individual of the future generations. Each of us is the descendant of a large population. The statement is not strictly true, for some lines die out, many lines cross, and caste has narrowed the field, but the statement suffices to show that a species moves along as a horde rather than as the offspring of a few individuals in each generation. The mass serves to keep the species afloat in times of calamity, it may have little else to do directly with its advance. Nevertheless this fundamental fact is too often overlooked in the attempt to explain the origin of new races, varieties and species from single favorable variations.

For *advance* we must look to those individuals that contribute something new to the species—it is the superman that will add something to the common level of humanity, but the rest keep the race alive until his advent and then carry his kind forward on an advancing wave.

If we could count those individuals that are the pioneers of advance, their number might be very small; in order to survive, they must graft themselves onto the stock. They are the harbingers of the better times to come—the forerunners of progress.

We touch here the crucial point of evolution in its relation to Darwin's principle of natural selection. Darwin says that he did not at first realize the overwhelming influence of the mass in its swamping effects on the individual variant. He made a very important concession to this view in the later editions of the "Origin of Species," and thought it necessary to assume that for a new form to arise it must first appear in a large number of individuals.

But to-day the situation has changed and new facts have come to light—facts that remove the enormous difficulty that Darwin met by what may seem now to have been an unnecessary concession.

An imaginary case will illustrate what I wish to say. Suppose that a species consisted in each generation of a million individuals and let us imagine that a new character—a definite variation—appears in an individual. The individual that bears it will pair with another ordinary individual and transmit its new character to all of its offspring. In order to simplify our case let us imagine that from each pair of individuals four reach maturity. The

million of individuals has increased to two millions, but accidents and competition may kill off one million of these, so that the race is again reduced to its standard of one million. If, then, we suppose that two of the new kinds of individuals survive on the average, and pair at random, there will be eight in the next generation (in reality only six of the eight will show this character). If these survive they will transmit their character to twelve of their offspring. Gradually, however, step by step, the new character will be added to the whole race. Thus any new, definite character will gradually appear in all the individuals whether it is useful or not. If it is useful it may sooner implant itself on the race than if it is indifferent; for more individuals may survive that possess it, than of those without it. It will spread faster, but in any case it will come in the long run. Thus we see that it spreads, not because it is advantageous, but because it is a definite variation. Injurious characters will have greater difficulties in inflicting themselves on the race, and if distinctly injurious may never succeed.

While one character is spreading, other definite variations may also be adding themselves to the race. Those individuals that combine the greatest number of useful additions will have the best chance of survival. Slowly the race advances in the direction of the sum of its advantages and adaptation; success, not in one but in several characters, is the true criterion of survival.

To fix our attention on each single advantage and to ascribe to it alone the palm of victory gives an incomplete idea of the progress of evolution, for evolution follows the line of the greatest number of adaptations. Success in every generation can not be traced to one variation, but to the sum of all mingled advantages.

This interpretation broadens, I think, our general conception of natural selection. We see that it is erroneous to suppose that all the individuals that bear a particular, useful trait owe this trait to their descent from one kind of individual, in the sense that this individual is the sole ancestor of all the later survivors. The first individual is not alone the ancestor of all the individuals that later bear its mark, it is only one of 999,999 ancestors that have contributed to the perpetuation of the race.

In order to simplify the case we have imagined that the new variation has appeared in a single individual. Should it appear in more than one, or arise again and again, its implantation would be thereby hastened, but the principle remains the same.

My contention may be summed up in a sentence. Survival value is not the only test for the perpetuation of any one useful character; it is the sum total of useful variations that determines progress. The species moves as a group always. Evolution is not a simple but a complex problem. This is the general opinion held by most modern zoologists.

To-day there are three great rival claims that attempt to explain how evolution takes place: (1) that which adopts the theory of natural selection in one or another of its aspects; (2) that which maintains that acquired characters are inherited; (3) that which, trying to penetrate deeper into the mystery of life, ascribes to living matter a purposefulness—an almost conscious response to "the course of nature."

In a few concluding words I shall try to point out the standing of these rival claims.

Darwin himself adopted both the first and the second of these views. His whole philosophy stands opposed in principle to the third view. He did not hesitate at times to adopt the theory of the inheritance of acquired characters, whenever the facts seemed more in accordance with that interpretation than with that of natural selection. He strenuously objected that he had never intended to refer the entire process of evolution to natural selection, and later in life affirmed that he had perhaps laid too little stress on the influence of the environment. To-day the doctrine of inherited effects is in disgrace, largely owing to the brilliant attack of the philosopher of Freiburg. Nevertheless it has warm adherents; and not a few of the most cautious zoologists now living have expressed themselves in its favor. It has not lacked able advocates, but it has sadly lacked direct evidence to support it. I can show you an example of how it fails when put to the test. I have here a waltzing mouse that turns round in circles instead of moving forwards. This is a domesticated variety and breeds true, i.e., all of its offspring are waltzers. I next show you a pair of mice that were injected with acetyl-atoxyl to cure them of sleeping sickness. They have artificially acquired the same habit as a result of the injection, and have waltzed for nearly a year. Here are their offspring that show not a trace of the trick.

Cases like these, and I could cite not a few, show how cautiously we must view the theory that such acquired characters are inherited. The experiments do not disprove the possibility, but

until direct evidence is forthcoming, judgment must remain suspended.

It has seemed, therefore, to many modern zoologists that we must face the two alternatives, either natural selection or purposeful response. Natural selection has been likened in recent years to a sieve that lets the non-adapted pass through and conserves the adapted. On the sieve metaphor natural selection produces nothing—it is described as a process of destruction of the unfit. How then can natural selection the destroyer become a factor in a creative process?

I have already tried to indicate how natural selection may assume such a rôle. If definite variations appear, however small or large, that are of some benefit, they may engraft themselves in time on to the species; if other useful definite variations are also adding themselves, if their presence insures some further definite variations in the same directions, advance is certain. In other words the elimination of the unfit has not produced the fit, but it has left the conditions more favorable for further progress in the direction of fitness. This is the interpretation of Darwinism that attracts at present the serious attention of the most thoughtful and advanced students of evolution.

I hesitate to bring before you in a closing sentence or two the alternative doctrine of purposefulness—a doctrine so fraught with human and superhuman import, for of all theories of creation it undoubtedly makes the strongest emotional appeal to mankind.

We are so conversant with the fact in human affairs that whenever purpose is involved there is an intelligent agent—a mind that designs, a mind that foresees—that our thinking has become tinctured with the idea that *wherever* there is purpose there is something like mind that has anticipated it. Organic nature is full to the brim of what seems purposeful adaptation—means to ends. Two modern zoologists and a noted philosopher have nailed this banner to their masthead.

There is one consideration above all others that warns the zoologist against speaking dogmatically about purposefulness, or its absence, in the response of living matter to its environment—his ignorance of the causes of variation. If I have implied that *all* variation is purely "accidental"; if I have led you to infer that it is entirely fortuitous, I have gone beyond the facts. We must be careful to distinguish between the individual differences that we can safely ascribe to chance, and the small definite variations

that arise in the germ. The latter appear to be limited, to be in part determined by the internal nature of the parts affected, and to be constant when they have once appeared, but more than this we dare not affirm. We may believe if we like that the evidence indicates that they are not purposeful, but we can not prove this. If *they* are not purposeful then the purposefulness of the living world has no *direct* relation to the origin of useful variations. The origin of an adaptive structure and the purpose it comes to fulfill are only chance combinations. Purposefulness is a very human conception for usefulness. It is usefulness looked at backwards. Hard as it is to imagine, inconceivably hard it may appear to many, that there is no direct relation between the origin of useful variations and the ends they come to serve, yet the modern zoologist takes his stand as a man of science on this ground. He may admit in secret to his father confessor, the metaphysician, that his poor intellect staggers under such a supposition, but he bravely carries forward his work of investigation along the only lines that he has found fruitful.

In the last analysis it is a matter of expediency; or if the word jars, a matter of instinct. Why forsake the gold mine at our feet, because the transmutation of metals is a philosophic possibility?

Whether definite variations are by chance useful, or whether they are purposeful are the contrasting views of modern speculation. The philosophic zoologist of to-day has made his choice. He has chosen *undirected* variations as furnishing the materials for natural selection. It gives him a working hypothesis that calls in no unknown agencies; it accords with what he observes in nature; it promises the largest rewards. He does not deny, if he is cautious, the possibility that there may be a purposefulness in the sense that organisms may respond adaptively at times to external conditions; for the very basis of his theory rests on the assumption that such variations do occur. But he is inclined to question the assumption that adaptive variations *arise because* of their adaptiveness. In his experience he finds little evidence for this belief, and he finds much that is opposed to it. He can foresee that to admit it for that all-important group of facts, where adjustments arise through the adaptation of individuals to each other—of host to parasite, of hunter to hunted—will land him in a mire of unverifiable speculation. He *fears* to enter thereby on a field of exploitation of nature that has proved itself so sterile in the past.

We have reached the end of our theme. If I have led you too far

into some of the remote corners of zoological thought, I must plead that such thoughts are the legitimate outgrowth of Darwinism.

I have tried to show you the modern zoologist at work on the great theory of evolution. We stand to-day on the foundations laid fifty years ago. Darwin's method is our method, the way he pointed out we follow, not as the advocates of a dogma, not as the disciples of any particular creed, but the avowed adherents of a method of investigation whose inauguration we owe chiefly to Charles Darwin. For it is this spirit of Darwinism, not its formulae, that we proclaim as our best heritage.

1909

Edward M. East

A Mendelian Interpretation of Variation That Is Apparently Continuous

> The major obstacle to the acceptance of Mendelism as a mechanism for evolution was the fluctuating variability of characters. The similarity of distribution of quantitative traits to a normal curve was the major reason why, at first, the biometric school rejected Mendelism. When Johannsen showed that selection was ineffective in inbred (pure or homozygous) lines, he thought the relevance of Mendelism would no longer be denied. The *genotype* of the individual was its genetic constitution; the *phenotype* of the individual was its appearance. But Johannsen did *not* show that the genotypes of his pure lines contained Mendelizing factors. E. M. East, however, using maize, and H. Nilsson-Ehle using wheat, demonstrated that quantitative traits could be produced through the joint activities of several genes, each one of which is insufficient to manifest the desired phenotype. Characters were no longer "unit characters" but complex resultants of the several participating genes. Multiple-factor inheritance was the chief conceptual means for associating Mendelism and evolution. It was this recognition that revived Darwinism, and the geneticists and evolutionists who were enthusiasts for this interpretation have been called Neo-Darwinists.

There are two objects in writing this paper.[1] One is to present some new facts of inheritance obtained from pedigree cultures of maize; the other is to discuss the hypotheses to which an extension of this class of facts naturally leads. This discussion is to be regarded simply as a suggestion toward a working hypothesis, for the facts are not sufficient to support a theory. They do, however, impose certain limitations upon speculation which should receive careful consideration.

Reprinted from *The American Naturalist 44*:65–82, 1910, by permission of the publisher.

The facts which are submitted have to do with independent allelomorphic pairs which cause the formation of like or similar characters in the zygote. Nilsson-Ehle[2] has just published facts of the same character obtained from cultures of oats and of wheat. My own work is largely supplementary to his, but it had been given these interpretations previous to the publication of his paper.

In brief, Nilsson-Ehle's results are as follows: He found that while in most varieties of oats with black glumes blackness behaved as a simple Mendelian mono-hybrid, yet in one case there were two definite independent Mendelian unit characters, each of which was allelomorphic to its absence. Furthermore, in most varieties of oats having a ligule, the character behaved as a mono-hybrid dominant to absence of ligule, but in one case no less than four independent characters for presence of ligule, each being dominant to its absence, were found. In wheat a similar phenomenon occurred. Many crosses were made between varieties having red seeds and those having white seeds. In every case but one the F_2 generation gave the ordinary ratio of three red to one white. In the one exception—a very old red variety from the north of Sweden—the ratio in the F_2 generation was 63 red to 1 white. The reds of the F_2 generation gave in the F_3 generation a very close approximation to the theoretical expectation, which is 37 constant red, 8 red and white separating in the ratio of 63:1, 12 red and white separating in the ratio of 15:1, 6 red and white separating in the ratio of 3:1, and one constant white. He did not happen to obtain the expected constant white, but in the total progeny of 78 F_2 plants his other results are so close to the theoretical calculation that they quite convince one that he was really dealing with three indistinguishable but independent red characters, each allelomorphic to its absence. Nor can the experimental proof of the two colors of the oat glumes be doubted. The evidence of four characters for presence of ligule in the oat is not so conclusive.

In my own work there is sufficient proof to show that in certain cases the endosperm of maize contains two indistinguishable, independent yellow colors, although in most yellow races only one color is present. There is also some evidence that there are three and possibly four independent red colors in the pericarp, and two colors in the aleurone cells. The colors in the aleurone cells when pure are easily distinguished, but when they are together they grade into each other very gradually.

Fully fifteen different yellow varieties of maize have been crossed with various white varieties, in which the crosses have all given a simple mono-hybrid ratio. In the other cases that follow it is seen that there is a di-hybrid ratio.

No. 5-20, a pure white eight-rowed flint, was pollinated by No. 6, a dent pure for yellow endosperm. An eight-rowed ear was obtained containing 159 medium yellow kernels and 145 light yellow kernels. The pollen parent was evidently a hybrid homozygous for one yellow which we will call Y_1 and heterozygous for another yellow Y_2. The gametes Y_1Y_2 and Y_1 fertilized the white in equal quantities, giving a ratio of approximately one medium yellow to one light yellow. The F_2 kernels from the dark yellow were as follows:

TABLE I[3]

F_2 SEEDS FROM CROSS OF NO. 5-20, WHITE FLINT × NO. 6 YELLOW DENT, HOMOZYGOUS FOR Y_1 AND HETEROZYGOUS FOR Y_2

Dark Seeds Heterozygous for Both Yellows Planted

Ear No.	Dark Y.	Light Y.	Total Y.	No Y.
1	270	56	326	29
2	101	215	316	27
3	261	52	313	28
5	273	284	557	35
10	358	117	475	25
12	296	72	368	19
13	207	156	363	35
14	387	102	489	29
Total	2153	1054	3207	227
Ratio			14.1	1

The ratios of light yellows to dark yellows is very arbitrary, for there was a fine gradation of shades. The ratio of total yellows to white, however, is unmistakably 15:1.

In the next table (Table II) are given the results of F_2 kernels from the light yellows of F_1. Only ear No. 8, which was really planted with the dark yellows, showed yellows dark enough to be mistaken for kernels containing both Y_1 and Y_2. The remain-

ing ears are clearly mono-hybrids with reference to yellow endosperm.

In a second case the female parent possessed the yellow endosperm. No. 11, a twelve-rowed yellow flint, was crossed with No. 8, a white dent. The F_2 kernels in part showed clearly a mono-hybrid ratio, and in part blended gradually into white. Two of these indefinite ears proved in the F_3 generation to have had the 15:1 ratio in the F_2 generation. Ear 7 of the F_2 generation calculated from the results of the entire F_3 crop must have had about 547 yellow to 52 white kernels, the theoretical number being 561 to 31. The hand-pollinated ears of the F_3 generation (yellow seeds) gave the results shown in Table III.

TABLE II

F_2 SEEDS FROM SAME CROSS AS SHOWN IN TABLE I

Light Yellow Seeds Heterozygous for Y_1 Planted

Ear No.	Dark Y.	Light Y.	No Y.
1		359	117
2		144	54
3		173	63
4		433	136
6		316	120
8	331		109
8a		229	86
9		325	115
10		227	87
11[4]		4	434
12		318	118
13		256	93
Total		3111	1098
Ratio		2.8	1

The F_3 generation grown from the other ear, Ear No. 8, showed that the ratio of yellows to whites in the F_2 generation was about 227 to 47. As the theoretical ratio is 257 to 17, the ratio obtained is somewhat inconclusive. A classification of the open field crop could not be made accurately on account of the light color of the yellows and the presence of many kernels showing zenia. Table IV,

however, showing the hand-pollinated kernels of the inter-bred yellows of the F_2 generation, settles beyond a doubt the fact that the two yellows were present.

TABLE III

NO. 11 YELLOW × NO. 8 WHITE

F_3 Generation from Yellow Seeds of F_2 Generation

Ear No.	Dark Y.	Light Y.	Total Y.	No Y.	Ratio They Approximate
1	116	95	211	19	15Y:1 no Y
14			88	5	15Y:1 no Y
5	181	122			$3Y_1Y_2$:1 $Y_{1 \text{ or } 2}$
4		253		68	3Y:1 no Y
6		193		73	"
8		163		79	"
11		108		35	"
9		456			Constant $Y_{1 \text{ or } 2}$

TABLE IV

PROGENY OF EAR NO. 8 OF THE SAME CROSS AS SHOWN IN TABLE III

F_3 Generation from Yellow Seeds of F_2 Generation

Ear No.	Dark Y.	Light Y.	Total Y.	No Y.	Ratio They Approximate
10	101	188	289	25	15Y:1 no Y
11	89	219	308	23	15Y:1 no Y
3		233			constant light Y
9	dark and light		331		3 dark:1 light Y
13	dark and light		350		3 dark:1 light Y
8		294		108	3 light:1 no Y
15		221		87	3 light:1 no Y
1[5]		197		203	

In a third case an eight-rowed yellow flint, No. 22, was crossed with a white dent, No. 8. Only four selfed ears were obtained in the F_2 generation. Ear 1 had 72 yellow to 37 white kernels. This ear was poorly developed and undoubtedly had some yellow kernels which were classed as whites. Ear 4 had 158 yellow and 42

white kernels. It is very likely that both of these ears were mono-hybrids, but the F_3 generation was not grown. Ear 5 had 148 yellow and 15 white kernels. Ear 7 had 78 yellow and 5 white kernels. It seems probable that both of these ears were di-hybrids, but only Ear 5 was grown another generation. The kernels classed as white proved to be pure; the open field crop from the yellow kernels gave 14 pure yellow ears and 14 hybrid yellow. Theoretically the ratio should be 7 pure yellows (that is, pure for either one or both yellows) and 8 hybrid yellows (4 giving 15 yellows to 1 white and 4 giving 3 yellows to 1 white). Five hand-pollinated selfed ears were obtained. Three of these gave mono-hybrid ratios, with a total of 607 yellows to 185 white kernels. One ear was a pure dark yellow (probably $Y_1Y_1Y_2Y_2$). The other ear was poorly filled, but had 27 dark yellows (probably Y_1Y_2) and 7 light yellow kernels (Y_1 or Y_2). Unfortunately no 15:1 ratio was obtained in this generation, but this is quite likely to happen when only five selfed ears are counted. The gradation of colors and the general appearance of the open field crop, however, lead me to believe that we were again dealing with a di-hybrid.

Two yellows appeared in still another case, that of white sweet No. 40 ♀ × yellow dent No. 3 ♂. Only one selfed ear was obtained in the F_2 generation giving 599 yellow to 43 white kernels. Of these kernels 486 were starchy and 156 sweet, which complicated matters in the F_3 generation because it was very difficult to separate the light yellow sweet from the white sweet kernels. Among the selfed ears were three pure to the starchy character, and in these ears the dark yellows, the light yellows and whites stood out very distinctly. Ear 12 had 156 dark yellow; 47 light yellow; 14 white kernels. Ear 13 had 347 dark yellow; 93 light yellow; 25 white kernels. The third starchy ear, No. 6 had 320 light yellow; 97 white kernels. Two ears, therefore, were di-hybrids, and one ear a mono-hybrid.

The ears which were heterozygous for starch and no starch and those homozygous for no starch, could not all be classified accurately, but it is certain that some pure dark yellows, some pure light yellows, some showing segregation of yellows and whites at the ratio 15:1, and some showing segregation of yellows and whites at the ratio of 3:1, were obtained.

One other case should be mentioned. One ear of a dent variety of unknown parentage obtained for another purpose was found to have some apparently heterozygous yellow kernels. Seven selfed

ears were obtained from them, of which two were pure yellow. The other five ears each gave the di-hybrid ratio. There was a total of 1906 yellow seeds to 181 white seeds, which is reasonably close to the expected ratio, 1956 yellow to 131 white.

It is to be regretted that I can present no other case of this class that has been fully worked out, although several other characters which I have under observation in both maize and tobacco seem likely to be included ultimately. Nevertheless, the fact that we have to deal with conditions of this kind in studying inheritance is established; granting only that they will be somewhat numerous, it opens up an entirely new outlook in the field of genetics.

In certain cases it would appear that we may have several allelomorphic pairs each of which is inherited independently of the others, and each of which is separately capable of forming the same character. When present in different numbers in different individuals, these units simply form quantitative differences. It may be objected that we do not know that two colors that appear the same physically are exactly the same chemically. That is true; but Nilsson-Ehle's case of several unit characters for presence of ligule in oats is certainly one where each of several Mendelian units forms exactly the same character. It may be that there is a kind of biological isomerism, in which, instead of molecules of the same formula having different physical properties, there are isomers capable of forming the same character, although, through difference in construction, they are not allelomorphic to each other. At least it is quite a probable supposition that through imperfections in the mechanism of heredity an individual possessing a certain character should give rise to different lines of descent so that in the F_n generation when individuals of these different lines are crossed, the character behaves as a di-hybrid instead of as a mono-hybrid. In other words, it is more probable that these units arise through variation in different individuals and are combined by hybridization, than that actually different structures for forming the same character arise in the same individual.

On the other hand, there is a possibility of an action just the opposite of this. Several of these quantitative units which produce the same character may become attached like a chemical radical and again behave as a single pair. Nilsson-Ehle gives one case which he does not attempt to explain, where the same cross gave a 4:1 ratio in one instance and 8.4:1 ratio in another instance. In

his other work characters always behaved the same way; that is, either as one pair, two pairs, three pairs, etc. In my work, the yellow endosperm of maize has behaved differently in the same strain, but it is probably because the yellow parent is homozygous for one yellow and heterozygous for the other. They were known to be pure for one yellow, but it would take a long series of crosses to prove purity in two yellows.

Let us now consider what is the concrete result of the interaction of several cumulative units affecting the same character. Where there is simple presence dominant to absence of a number *n* of such factors, in a cross where all are present in one parent and all absent in the other parent, there must be 4^n individuals to run an even chance of obtaining a single F_2 individual in which the character is absent. When four such units, $A_1A_2A_3A_4$ are crossed with $a_1a_2a_3a_4$, their absence, only one pure recessive is expected in 256 individuals. And 256 individuals is a larger number than is usually reported in genetic publications. When a smaller population is considered, it will appear to be a blend of the two parents with a fluctuating variability on each side of its mode. Of course if there is absolute dominance and each unit appears to affect the zygote in the same manner that they do when combined, the F_2 generation will appear like the dominant parent unless a very large number of progeny are under observation and pure recessives are obtained. This may be an explanation of the results obtained by Millardet; it is certainly as probable as the hypothesis of the non-formation of homozygotes. Ordinarily, however, there is not perfect dominance, and variation due to heterozygosis combined with fluctuating variation makes it almost impossible to classify the individuals except by breeding. The two yellows in the endosperm of maize is an example of how few characters are necessary to make classification difficult. First, there is a small amount of fluctuation in different ears due to varying light conditions owing to differences in thickness of the husk; second, all the classes having different gametic formulae differ in the intensity of their yellow in the following order, $Y_1Y_1Y_2Y_2$, $Y_1y_1Y_2Y_2$ or $Y_1Y_1Y_2y_2$, Y_1Y_1, Y_2Y_2, Y_1y_1, Y_2y_2, y_1y_2. As dominance becomes less and less evident, the Mendelian classes vary more and more from the formula $(3 + 1)^n$, and approach the normal curve, with a regular gradation of individuals on each side of the mode. When there is no dominance and open fertilization, a state is reached in which the curve of variation simulates the

fluctuation curve, with the difference that the gradations are heritable.

One other important feature of this class of genetic facts must be considered. If units $A_1A_2A_3a_4$ meet units $a_1a_2a_3A_4$, in the F_2 generation there will be one pure recessive, $a_1a_2a_3a_4$, in every 256 individuals. This explains an apparent paradox. Two individuals are crossed, both seemingly pure for presence of the same character, yet one individual out of 256 is a pure recessive. When we consider the rarity with which pure dominants or pure recessives (for all characters) are obtained when there are more than three factors, we can hardly avoid the suspicion that here is a perfectly logical way of accounting for many cases of so-called atavism. Furthermore, many apparently new characters may be formed by the gradual dropping of these cumulative factors without any additional hypothesis. For example, in *Nicotiana tabacum* varieties there is every gradation[6] of loss of leaf surface near the base of the sessile leaf, until in *N. tabacum fruticosa* the leaf is only one step removed from a petioled condition. If this step should occur the new plant would almost certainly be called a new species; yet it is only one degree further in a definite series of loss gradations that have already taken place. If it should be assumed that in other instances slight qualitative as well as quantitative changes take place as units are added, then it becomes very easy, theoretically, to account for quite different characters in the individual homozygous for presence of all dominant units, and in the individual in which they are all absent.

Unfortunately for these conceptions, although I feel it extremely probable that variations in *some* characters that seem to be continuous will prove to be combinations of segregating characters, it is exceedingly difficult to demonstrate the matter beyond a reasonable doubt. As an illustration of the difficulties involved in the analysis of pedigree cultures embracing such characters, I wish to discuss some data regarding the inheritance of the number of rows of kernels on the maize cob.

The maize ear may be regarded as a fusion of four or more spikes, each joint of the rachis bearing two spikelets. The rows are, therefore, distinctly paired, and no case is known where one of the pair has been aborted. This is a peculiar fact when we consider the great number of odd kinds of variations that occur in nature. The number of rows per cob has been considered to belong to continuous variations by DeVries, and a glance at the

progeny from the seeds of a single selfed ear as shown in Table V seems to confirm this view.

There is considerable evidence, however, that this character is made up of a series of cumulative units, independent in their inheritance. There is no reason why it should not be considered to be of the same nature as various other size characters in which variation seems to be continuous, but in which relatively constant gradations may be isolated, each fluctuating around a particular mode. But this particular case possesses an advantage not held by most phenomena of its class, in that there is a definite discontinuous series of numbers by which each individual may be classified.

TABLE V

PROGENY OF A SELFED EAR OF LEAMING MAIZE HAVING 20 ROWS

Classes of rows	12	14	16	18	20	22	24	26	28	30
No. of ears	1	0	5	4	53	35	19	5	2	1

Previous to analyzing the data from pedigree cultures, however, it is necessary to take into consideration several facts. In the first place, what limits are to be placed on fluctuations?[7] From the variability of the progeny of single ears of dent varieties that have been inbred for several generations, it might be concluded that the deviations are very large. But this is not necessarily the case; these deviations may be due largely to gametic structure in spite of the inbreeding, since no conscious selection of homozygotes has been made. There is no such variation in eight-rowed varieties, which may be considered as the last subtraction form in which maize appears and therefore an extreme homozygous recessive. In a count of the population of an isolated maize field where Longfellow, an eight-rowed flint, had been grown for many years, 4 four-rowed, 993 eight-rowed, 2 ten-rowed and 1 twelve-rowed ears were found. Only seven aberrant ears out of a thousand had been produced, and some of these may have been due to vicinism.

On the other hand a large number of counts of the number of rows of both ears on stalks that bore two ears has shown that it is very rare that there is a change greater than ± 2 rows. If conditions are more favorable at the time when the upper ear is laid down it will have two more rows than the second ear; if conditions

are favorable all through the season, the ears generally have the same number of rows; while if conditions are unfavorable when the upper ear is laid down, the lower ear may have two more rows than the upper ear. Furthermore, seeds from the same ear have several times been grown on different soils and in different seasons, and in each case the frequency distribution has been the same. Hence it may be concluded that in the great majority of cases fluctuation is not greater than in ± 2 rows, although fluctuations of ± 4 rows have been found.

A second question worthy of consideration is: Do somatic variations due to varying conditions during development take place with equal frequency in individuals with a large number of rows and in individuals with a small number of rows? From the fact that several of my inbred strains that have been selected for three generations for a constant number of rows, increase directly in variability as the number of rows increases, the question should probably be answered in the negative. This answer is reasonable upon other grounds. The eight-rowed ear may vary in any one of four spikes, the sixteen-rowed ear may vary in any one of eight spikes; therefore the sixteen-rowed ear may vary twice as often as the eight-rowed ear. By the same reasoning, the sixteen-rowed ear may sometimes throw fluctuations twice as wide as the eight-rowed ear.

A third consideration is the possibility of increased fluctuation due to hybridization. Shull[8] and East[9] have shown that there is an increased stimulus to cell division when maize biotypes are crossed—a phenomenon apart from inheritance. There is no evidence, however, that increased gametic variability results. Johannsen[10] has shown that there is no such increase in fluctuation when close-pollinated plants are crossed. I have crossed several distinct varieties of maize where the modal number of rows of each parent was twelve, and in every instance the F_1 progeny had the same mode and about the same variability.

Finally, a possibility of gametic coupling should be considered. Our common races of lint maize all have a low number of rows, usually eight but sometimes twelve; dent races have various modes running from twelve to twenty-four rows. When crosses between the two sub-species are made, the tendency is to separate in the same manner.

Attention is not called to these obscuring factors with the idea that they are universally applicable in the study of supposed con-

tinuous variation. But there are similar conditions always present that make analysis of these variations difficult, and the facts given here should serve to prevent premature decision that they do not show segregation in their inheritance.

Table VI shows the results from several crosses between maize races with different modal values for number of rows. Several interesting points are noticeable. The modal number is always divisible by four. This is also the case with some twenty-five other races that I have examined but which are not shown in the table. I suspect that through the presence of pure units zygotes having a multiple of four rows are formed, while heterozygous units cause the dropping of two rows. The eight-rowed races are pure for that character, the twelve-rowed races vary but little, but the races having a higher number of rows are exceedingly variable.

When twelve-rowed races are crossed with those having eight rows, the resulting F_1 generation always—or nearly always—has the mode at twelve rows. In one case cited in Table VI, No. 24 × No. 53, nearly all the F_1 progeny were eight-rowed. It might appear from this, either that the low number of rows was in this case dominant, or that the female parent has more influence on the resulting progeny than the male parent. I prefer to believe, however, that the individual of No. 53 which furnished the pollen was due to produce eight-rowed progeny. Unfortunately no record was kept of the ear borne by this plant, but No. 53 sometimes does produce eight-rowed ears.

When a race with a mode higher than twelve is crossed with an eight-rowed race, the F_1 generation is always intermediate, although it tends to be nearer the high-rowed parent. Only one example is given in the table, but it is indicative of the class. These results are rather confusing, for there seems to be a tendency to dominance in the twelve-rowed form that is not found in the forms with a higher number of rows. I have seen cultures of other investigators where 12-row × 8-row resulted in a ten-rowed F_1 generation, so the complication need not worry us at present.

The results of the F_2 generation show a definite tendency toward segregation and reproduction of the parent types. I might add that in at least two cases I have planted extracted eight-rowed ears and have immediately obtained an eight-rowed race which showed only slight departures from the type. Selection from those

TABLE VI

CROSSES BETWEEN MAIZE STRAINS WITH DIFFERENT NUMBERS OF ROWS

Parents (Female Given First)	Gen.	Row Classes						
		8	10	12	14	16	18	20
Flint No. 5		100						
Flint No. 11		1	4	387	7	1		
Flint No. 24		100						
Flint No. 15		100						
Dent No. 6				6	31	51	18	4
Dent No. 8			3	54	36	12	2	
Sweet No. 53[11]		1	5	25	4			
Sweet No. 54[11]		25	2	1				
No. 5 × No. 53	F_1	1	7	13				
No. 5 × No. 6	F_1	11	18	27	3			
No. 11 × No. 5	F_1	2	4	18				
No. 11 × No. 53	F_1	2	5	17				
No. 24 × No. 53	F_1	57	8	3				
No. 15 × No. 8	F_1	1	14	26	3	1		
No. 15 × No. 8 (from 10-row ear)	F_2	14	15	28	9	1		
No. 15 × No. 8 (from 12-row ear)	F_2	4	13	25	6	3		
No. 8 × No. 54	F_1	1	6	14				
No. 8 × No. 54 (from 12-row ear)	F_2	11	25	38	2	1		

ears having a high number of rows has also given races like the high-rowed parent without recrossing with it. It is regretted that commercial problems were on hand at the time and no exact data were recorded. It can be stated with confidence, however, that ears like each parent are obtained in the F_2 generation, from which with care *races* like each parent may be produced. *Segregation seems to be the best interpretation of the matter.*

These various items may seem disconnected and uninteresting, but they have been given to show the tangible basis for the following theoretical interpretation. No hard and fast conclusion is attempted, but I feel that this interpretation with possibly slight modifications will be found to aid the explanation of many cases where variation is apparently continuous.

Suppose a basal unit to be present in the gametes of all maize

races, this unit to account for the production of eight rows. Let additional independent interchangeable units, each allelomorphic to its own absence, account for each additional four rows; and let the heterozygous condition of any unit represent only half of the homozygous condition, or two rows. Then the gametic condition of a homozygous twenty-rowed race would be 8 + AABBCC, each letter actually representing two rows. When crossed with an eight-rowed race, the F_2 generation will show ears of from eight to twenty rows, each class being represented by the number of units in the coefficients in the binomial expansion where the exponent is twice the number of characters, or in this case $(a + b)^6$.

The result appears to be a blend between the characters of the two parents with a normal frequency distribution of the deviants. Only one twenty-rowed individual occurs in 64 instead of the 27 expected by the interaction of three dominant factors in the usual Mendelian ratios. The remainder of the 27 will have different numbers of rows, and, by their gametic formulae, different expectations in future breeding as follows:

1 AABBCC = 20 rows.
2 AaBBCC = 18 rows.
2 AABbCC = 18 rows.
2 AABBCc = 18 rows.
4 AaBbCC = 16 rows.
4 AaBBCc = 16 rows.
4 AABbCc = 16 rows.
8 AaBbCc = 14 rows.

There are four visibly different classes and eight gametically different classes. It must also be remembered that the probability that the original twenty-rowed ear in actual practice may have had more than three units in its gametes has not been considered. This point is illustrated clearly if we work out the complete ratio for the three characters, and note the number of gametically different classes which compose the modal class of fourteen rows in Table VII.

It actually contains seven gametically different classes and not a single homozygote. If this conception of independent allelomorphic pairs affecting the same character proves true, it will sadly upset the biometric belief that the modal class is *the type* around which the variants converge, for there is actually less

chance of these individuals breeding true than those from *any other* class.

The conception is simple and is capable theoretically of bringing in order many complicated facts, although the presence of fluctuating variation will be a great factor in preventing analysis of data. I have thought of only one fact that is difficult to bring into line. If 8AA, 8BB and 8CC all represent homozygous twelve-rowed ears—to continue the maize illustration—and none of these factors are allelomorphic to each other, sixteen-rowed ears should sometimes be obtained when crossing two twelve-rowed ears. I am not sure but that this would happen if we were to extract all the homozygous twelve-rowed strains after a cross between sixteen-row and eight-row, and after proving their purity cross them. In some cases the additional four-row units would probably be allelomorphic to each other and in other cases independent of each other. On the other hand, this is only an hypothesis, and while I have faith in its foundation facts, the details may need change.

TABLE VII

THEORETICAL EXPECTATION IN F_2 WHEN A HOMOZYGOUS TWENTY-ROWED MAIZE EAR IS CROSSED WITH AN EIGHT-ROWED EAR

Classes . . .	8	10	12	14	16	18	20
No. ears . .	1	6	15	20	15	6	1

Castle has raised the point that greater variation should be expected in the F_1 generation than in the P_1 generations when crossing widely deviating individuals showing variation apparently continuous. If the parents are strictly pure for a definite number of units, say for size, a greater variation should certainly be expected in the F_1 generation after crossing. But considering the difficulties that arise when even five independent units are considered, can it be said that anything has heretofore been known concerning the actual gametic status of parents which it is known do vary in the character in question and in which the variations are inherited, for the race can be changed by selection within it. It may be, too, that the correct criterion has not been used in size measurements, for, as others have suggested, solids

vary as the cube root of their mass, whereas the sum of the weights of the body cells has usually been measured and compared directly with similar sums.

Attention should be called to one further point. Many characters in all probability are truly blending in their inheritance, but there is another interpretation which may apply in certain cases. I have repeatedly tried to cross Giant Missouri Cob Pipe maize (14 feet high) and Tom Thumb pop maize (2 feet high), but have always failed. They both cross readily with varieties intermediate in size, but are sterile between themselves. We may imagine that the gametes of each race, though varying in structure, are all so dissimilar that none of them can unite to form zygotes. Other races may be found where only part of the gametes of varying structure are so unlike that they will not develop after fusion. The zygotes that do develop will be from those more alike in construction. An apparent blend results, and although segregation may take place, no progeny as extreme as either of the parents will ever occur.

I may say in conclusion that the effect of the truth of this hypothesis would be to add another link to the increasing chain of evidence that the word "mutation" may properly be applied to any inherited variation, however small; and the word fluctuation should be restricted to those variations due to immediate environment which do not affect the germ cells, and which—it has been shown—are not inherited. In addition it gives a rational basis for the origin of *new* characters, which has hitherto been somewhat of a Mendelian stumbling-block; and also gives the term unit-character less of an irrevocably-fixed-entity conception, which is more in accord with other biological beliefs.

1910

NOTES

[1] Contributions from the Laboratory of Genetics, Bussey Institution, Harvard University, No. 4. Read before the annual meeting of the American Society of Naturalists, Boston, December 29, 1909.

[2] Nilsson-Ehle, H. *Kreuzungsuntersuchungen an Hafer und Weizen.* Lunds Universitets Årsskrift, N.F. Afd. 2, Bd. 5, No. 2, 1909.

[3] In these tables only hand pollinated ears are given.

[4] Discarded from average. This ear evidently grew from one kernel of the original white mother that was accidentally self-pollinated. The four yellow kernels all show zenia from accidental pollination in the next generation.

[5] Kernel from which this ear grew was evidently pollinated by no Y.

[6] It is not known at present how this character behaves in inheritance.

[7] The word fluctuation is used to designate the somatic changes due to immediate environment, and which *are not inherited.*

[8] Shull, G. H., "A Pure-line Method in Corn Breeding," *Rept. Amer. Breeders' Assn.,* 5, 51–59, 1909.

[9] East, E. M., "The Distinction between Development and Heredity in Inbreeding," *Amer. Nat.,* 43, 173–181, 1909.

[10] Johannsen, W., "Does Hybridization Increase Fluctuating Variability?" *Rept. Third Inter. Con. on Genetics,* 98–113, London, Spottiswoode, 1907.

[11] Approximately.

H. J. Muller

Variation Due to Change in the Individual Gene

This is one of the most prophetic articles written in biology. Muller emphasizes the importance of the individual gene. In reviewing its properties of replication, mutation, extreme stability, and vast physiological influence, Muller asserts the unique importance of genes as molecules. He suggests how gene size can be estimated; he points out the importance of mutation studies and the need for artificially changing the gene. The evolutionary significance of the gene he succinctly summarizes: "Thus it is not inheritance *and* variation which bring about evolution, but the inheritance *of* variation. . . ." Most remarkable is Muller's prediction that someday the gene will be studied biochemically with a union of genetics, zoology, botany, pathology, chemistry, and physics. His vision was based on the similarity of viruses (d'Hérelle bodies or bacteriophage) to genes themselves.

I. THE RELATION BETWEEN THE GENES AND THE CHARACTERS OF THE ORGANISM

The present paper[1] will be concerned rather with problems, and the possible means of attacking them, than with the details of cases and data. The opening up of these new problems is due to the fundamental contribution which genetics has made to cell physiology within the last decade. This contribution, which has so far scarcely been assimilated by the general physiologists themselves, consists in the demonstration that, besides the ordinary proteins, carbohydrates, lipoids, and extractives, of their several types, there are present within the cell *thousands* of distinct substances—the "genes"; these genes exist as ultra-microscopic par-

Reprinted from *The American Naturalist LVI,* Jan.–Feb., 1922, by permission of the publisher and the author.

ticles; their influences nevertheless permeate the entire cell, and they play a fundamental rôle in determining the nature of all cell substances, cell structures, and cell activities. Through these cell effects, in turn, the genes affect the entire organism.

It is not mere guesswork to say that the genes are ultra-microscopic bodies. For the work on *Drosophila* has not only proved that the genes are in the chromosomes, in definite positions, but it has shown that there must be hundreds of such genes within each of the larger chromosomes, although the length of these chromosomes is not over a few microns. If, then, we divide the size of the chromosome by the minimum number of its genes, we find that the latter are particles too small to give a visible image.

The chemical composition of the genes, and the formulae of their reactions, remain as yet quite unknown. We do know, for example, that in certain cases a given pair of genes will determine the existence of a particular enzyme (concerned in pigment production), that another pair of genes will determine whether or not a certain agglutinin shall exist in the blood, a third pair will determine whether homogentisic acid is secreted into the urine ("alkaptonuria"), and so forth. But it would be absurd, in the third case, to conclude that on this account the gene itself consists of homogentisic acid, or any related substance, and it would be similarly absurd, therefore, to regard cases of the former kind as giving any evidence that the gene *is* an enzyme, or an agglutinin-like body. The reactions whereby the genes produce their ultimate effects are too complex for such inferences. Each of these effects, which we call a "character" of the organism, is the product of a highly complex, intricate, and delicately balanced system of reactions, caused by the interaction of countless genes, and every organic structure and activity is therefore liable to become increased, diminished, abolished, or altered in some other way, when the balance of the reaction system is disturbed by an alteration in the nature or the relative quantities of any of the component genes of the system. To return now to these genes themselves.

II. THE PROBLEM OF GENE MUTABILITY

The most distinctive characteristic of each of these ultra-microscopic particles—that characteristic whereby we identify it as a gene—is its property of self-propagation: the fact that, within the

complicated environment of the cell protoplasm, it reacts in such a way as to convert some of the common surrounding material into an end-product identical in kind with the original gene itself. This action fulfills the chemist's definition of "autocatalysis"; it is what the physiologist would call "growth"; and when it passes through more than one generation it becomes "heredity." It may be observed that this reaction is in each instance a rather highly localized one, since the new material is laid down by the side of the original gene.

The fact that the genes have this autocatalytic power is in itself sufficiently striking, for they are undoubtedly complex substances, and it is difficult to understand by what strange coincidence of chemistry a gene can happen to have just that very special series of physico-chemical effects upon its surroundings which produces —of all possible end-products—just this particular one, which is identical with its own complex structure. But the most remarkable feature of the situation is not this oft-noted autocatalytic action in itself—it is the fact that, when the structure of the gene becomes changed, through some "chance variation," the catalytic property of the gene may[2] become correspondingly changed, in such a way as to leave it still autocatalytic. In other words, the change in gene structure—accidental though it was—has somehow resulted in a change of exactly *appropriate* nature in the catalytic reactions, so that the new reactions are now accurately adapted to produce more material just like that in the new changed gene itself. It is this paradoxical phenomenon which is implied in the expression "variation due to change in the individual gene," or, as it is often called, "mutation."

What sort of structure must the gene possess to permit it to mutate in this way? Since, through change after change in the gene, this same phenomenon persists, it is evident that it must depend upon some general feature of gene construction—common to all genes—which gives each one a *general* autocatalytic power—a "carte blanche"—to build material of whatever specific sort it itself happens to be composed of. This general principle of gene structure might, on the one hand, mean nothing more than the possession by each gene of some very simple character, such as a particular radicle or "side-chain"—alike in them all—which enables each gene to enter into combination with certain highly organized materials in the outer protoplasm, in such a way as to result in the formation, "by" the protoplasm, of more material

like this gene which is in combination with it. In that case the gene itself would only initiate and guide the direction of the reaction. On the other hand, the extreme alternative to such a conception has been generally assumed, perhaps gratuitously, in nearly all previous theories concerning hereditary units; this postulates that the chief feature of the autocatalytic mechanism resides in the structure of the genes themselves, and that the outer protoplasm does little more than provide the building material. In either case, the question as to what the general principle of gene construction is, that permits this phenomenon of mutable autocatalysis, is the most fundamental question of genetics.

The subject of gene variation is an important one, however, not only on account of the apparent problem that is thus inherent in it, but also because this same peculiar phenomenon that it involves lies at the root of organic evolution, and hence of all the vital phenomena which have resulted from evolution. It is commonly said that evolution rests upon two foundations—inheritance and variation; but there is a subtle and important error here. Inheritance by itself leads to no change, and variation leads to no permanent change, unless the variations themselves are heritable. Thus it is not inheritance *and* variation which bring about evolution, but the inheritance *of* variation, and this in turn is due to the general principle of gene construction which causes the persistence of autocatalysis despite the alteration in structure of the gene itself. Given, now, any material or collection of materials having this one unusual characteristic, and evolution would automatically follow, for this material would, after a time, through the accumulation, competition and selective spreading of the self-propagated variations, come to differ from ordinary inorganic matter in innumerable respects, in addition to the original difference in its mode of catalysis. There would thus result a wide gap between this matter and other matter, which would keep growing wider, with the increasing complexity, diversity and so-called "adaptation" of the selected mutable material.

III. A POSSIBLE ATTACK THROUGH CHROMOSOME BEHAVIOR

In thus recognizing the nature and the importance of the problem involved in gene mutability have we now entered into a *cul de sac,* or is there some way of proceeding further so as to get at the

physical basis of this peculiar property of the gene? The problems of growth, variation and related processes seemed difficult enough to attack even when we thought of them as inherent in the organism as a whole or the cell as a whole—how now can we get at them when they have been driven back, to some extent at least, within the limits of an invisible particle? A gene cannot effectively be ground in a mortar, or distilled in a retort, and although the physico-chemical investigation of other biological substances may conceivably help us, by analogy, to understand its structure, there seems at present no method of approach along this line.

There is, however, another possible method of approach available: that is, to study the behavior of the chromosomes, as influenced by their contained genes, in their various physical reactions of segregation, crossing over, division, synapsis, etc. This may at first sight seem very remote from the problem of getting at the structural principle that allows mutability in the gene, but I am inclined to think that such studies of synaptic attraction between chromosomes may be especially enlightening in this connection, because the most remarkable thing we know about genes—besides their mutable autocatalytic power—is the highly specific attraction which like genes (or local products formed by them) show for each other. As in the case of the autocatalytic forces, so here the attractive forces of the gene are somehow exactly adjusted so as to react in relation to more material of the same complicated kind. Moreover, when the gene mutates, the forces become readjusted, so that they may now attract material of the new kind; this shows that the attractive or synaptic property of the gene, as well as its catalytic property, is not primarily dependent on its specific structure, but on some general principle of its make-up, that causes whatever specific structure it has to be auto-attractive (and autocatalytic).

This auto-attraction is evidently a strong force, exerting an appreciable effect against the non-specific mutual repulsions of the chromosomes, over measurable microscopic distances much larger than in the case of the ordinary forces of so-called cohesion, adhesion and adsorption known to physical science. In this sense, then, the physicist has no parallel for this force. There seems, however, to be no way of escaping the conclusion that in the last analysis it must be of the same nature as these other forces which cause inorganic substances to have specific attractions for each

other, according to their chemical composition. These inorganic forces, according to the newer physics, depend upon the arrangement and mode of motion of the electrons constituting the molecules, which set up electro-magnetic fields of force of specific patterns. To find the principle peculiar to the construction of the force-field pattern of genes would accordingly be requisite for solving the problem of their tremendous auto-attraction.

Now, according to Troland (1917), the growth of crystals from a solution is due to an attraction between the solid crystal and the molecules in solution caused by the similarity of their force field patterns, somewhat as similarly shaped magnets might attract each other—north to south poles—and Troland maintains that essentially the same mechanism must operate in the autocatalysis of the hereditary particles. If he is right, each different portion of the gene structure must—like a crystal—attract to itself from the protoplasm materials of a similar kind, thus moulding next to the original gene another structure with similar parts, identically arranged, which then become bound together to form another gene, a replica of the first. This does not solve the question of what the general principle of gene construction is, which permits it to retain, like a crystal, these properties of auto-attraction,[3] but if the main point is correct, that the autocatalysis is an expression of specific attractions between portions of the gene and similar protoplasmic building blocks (dependent on their force-field patterns), it is evident that the very same forces which cause the genes to grow should also cause like genes to attract each other, but much more strongly, since here all the individual attractive forces of the different parts of the gene are summated. If the two phenomena are thus really dependent on a common principle in the make-up of the gene, progress made in the study of one of them should help in the solution of the other.

Great opportunities are now open for the study of the nature of the synaptic attraction, especially through the discovery of various races having abnormal numbers of chromosomes. Here we have already the finding by Belling, that where three like chromosomes are present, the close union of any two tends to exclude their close union with the third. This is very suggestive, because the same thing is found in the cases of specific attractions between inorganic particles, that are due to their force-field patterns. And through Bridges' finding of triploid *Drosophila*, the attraction phenomena can now be brought down to a definitely genic basis,

by the introduction of specific genes—especially those known to influence chromosome behavior—into one of the chromosomes of a triad. The amount of influence of this gene on attraction may then be tested quantitatively, by genetic determination of the frequencies of the various possible types of segregation. By extending such studies to include the effect of various conditions of the environment—such as temperature, electrostatic stresses, etc.—in the presence of the different genetic situations, a considerable field is opened up.

This suggested connection between chromosome behavior and gene structure is as yet, however, only a possibility. It must not be forgotten that at present we cannot be sure that the synaptic attraction is exerted by the genes themselves rather than by local products of them, and it is also problematical whether the chief part of the mechanism of autocatalysis resides within the genes rather than in the "protoplasm." Meanwhile, the method is worth following up, simply because it is one of our few conceivable modes of approach to an all-important problem.

It may also be recalled in this connection that besides the genes in the chromosomes there is at least one similarly autocatalytic material in the chloroplastids, which likewise may become permanently changed, or else lost, as has been shown by various studies on chlorophyll inheritance. Whether this plastid substance is similar to the genes in the chromosomes we cannot say, but of course it cannot be seen to show synaptic attraction, and could not be studied by the method suggested above.[4]

IV. THE ATTACK THROUGH STUDIES OF MUTATION

There is, however, another method of attack, in a sense more direct, and not open to the above criticisms. That is the method of investigating the individual gene, and the structure that permits it to change, through a study of the changes themselves that occur in it, as observed by the test of breeding and development. It was through the investigation of the *changes* in the chromosomes—caused by crossing over—that the structure of the chromosomes was analyzed into their constituent genes in line formation; it was through study of molecular changes that molecules were analyzed into atoms tied together in definite ways, and it has been finally the rather recent finding of changes in atoms and investigation of the resulting pieces, that has led us

to the present analysis of atomic structure into positive and negative electrons having characteristic arrangements. Similarly, to understand the properties and possibilities of the individual gene, we must study the mutations as directly as possible, and bring the results to bear upon our problem.

(a) *The Quality and Quantity of the Change*

In spite of the fact that the drawing of inferences concerning the gene is very much hindered, in this method, on account of the remoteness of the gene-cause from its character-effect, one salient point stands out already. It is that the change is not always a mere loss of material, because clear-cut reverse mutations have been obtained in corn, *Drosophila, Portulaca,* and probably elsewhere. If the original mutation was a loss, the reverse must be a gain. Secondly, the mutations in many cases seem not to be quantitative at all, since the different allelomorphs formed by mutations of one original gene often fail to form a single linear series. One case, in fact, is known in which the allelomorphs even affect totally different characters: this is the case of the truncate series, in which I have found that different mutant genes at the same locus may cause either a shortening of the wing, an eruption on the thorax, a lethal effect or any combination of two or three of these characters. In such a case we may be dealing either with changes of different types occurring in the same material or with changes (possibly quantitative changes, similar in type) occurring in different component parts of one gene. Owing to the universal applicability of the latter interpretation, even where allelomorphs do not form a linear series, it cannot be categorically denied, in any individual case, that the changes may be merely quantitative changes of some *part* of the gene. If all changes were thus quantitative, even in this limited sense of a loss or gain of part of the gene, our problem of why the changed gene still seems to be autocatalytic would in the main disappear, but such a situation is excluded a priori since in that case the thousands of genes now existing could never have evolved.

Although a given gene may thus change in various ways, it is important to note that there is a strong tendency for any given gene to have its changes of a particular kind, and to mutate in one direction rather than in another. And although mutation certainly does not always consist of loss, it often gives effects that

might be termed losses. In the case of the mutant genes for bent and eyeless in the fourth chromosome of *Drosophila* it has even been proved, by Bridges, that the effects are of exactly the same kind, although of lesser intensity, than those produced by the entire loss of the chromosome in which they lie, for flies having bent or eyeless in one chromosome and lacking the homologous chromosome are even more bent, or more eyeless, than those having a homologous chromosome that also contains the gene in question. The fact that mutations are usually recessive might be taken as pointing in the same direction, since it has been found in several cases that the loss of genes—as evidenced by the absence of an entire chromosome of one pair—tends to be much more nearly recessive than dominant in its effect.

The effect of mutations in causing a loss in the characters of the organism should, however, be sharply distinguished from the question of whether the gene has undergone any loss. It is generally true that mutations are much more apt to cause an apparent loss in character than a gain, but the obvious explanation for that is, not because the gene tends to lose something, but because most characters require for proper development a nicely adjusted train of processes, and so any change in the genes—no matter whether loss, gain, substitution or rearrangement—is more likely to throw the developmental mechanism out of gear, and give a "weaker" result, than to intensify it. For this reason, too, the most frequent kind of mutation of all is the lethal, which leads to the loss of the entire organism, but we do not conclude from this that all the genes had been lost at the time of the mutation. The explanation for this tendency for most changes to be degenerative, and also for the fact that certain other kinds of changes—like that from red to pink eye in *Drosophila*—are more frequent than others—such as red to brown or green eye—lies rather in developmental mechanics than in genetics. It is because the developmental processes are more unstable in one direction than another, and easier to push "downhill" than up, and so any mutations that occur—no matter what the gene change is like—are more apt to have these *effects* than the other *effects*. If now selection is removed in regard to any particular character, these character changes which occur more readily must accumulate, giving apparent orthogenesis, disappearance of unused organs, of unused physiological capabilities, and so forth. As we shall see later, however, the changes are not so frequent or numerous that

they could ordinarily push evolution in such a direction against selection and against the immediate interests of the organism.

In regard to the magnitude of the somatic effect produced by the gene variation, the *Drosophila* results show that there the smaller character changes occur oftener than large ones. The reason for this is again probably to be found in developmental mechanics, owing to the fact that there are usually more genes slightly affecting a given character than those playing an essential role in its formation. The evidence proves that there are still more genes whose change does not affect the given character at all —no matter what this character may be, unless it is life itself— and this raises the question as to how many mutations are absolutely unnoticed, affecting no character, or no detectable character, to any appreciable extent at all. Certainly there must be many such mutations, judging by the frequency with which "modifying factors" arise, which produce an effect only in the presence of a special genetic complex not ordinarily present.

(b) *The Localization of the Change*

Certain evidence concerning the causation of mutations has also been obtained by studying the relations of their occurrence to one another. Hitherto it has nearly always been found that only one mutation has occurred at a time, restricted to a single gene in the cell. I must omit from consideration here the two interesting cases of deficiency, found by Bridges and by Mohr, in each of which it seems certain that an entire region of a chromosome, with its whole cargo of genes, changed or was lost, and also a certain peculiar case, not yet cleared up, which has recently been reported by Nilsson-Ehle; these important cases stand alone. Aside from them, there are only two instances in which two (or more) new mutant genes have been proved to have been present in the same gamete. Both of these are cases in *Drosophila*— reported by Muller and Altenburg (1921)—in which a gamete contained two new sex-linked lethals; two cases are not a greater number than was to have been expected from a random distribution of mutations, judging by the frequency with which single mutant lethals were found in the same experiments. Ordinarily, then, the event that causes the mutation is specific, affecting just one particular kind of gene of all the thousands present in the cell. That this specificity is due to a spatial limitation rather than

a chemical one is shown by the fact that when the single gene changes, the other one, of identical composition, located near by in the homologous chromosome of the same cell, remains unaffected. This has been proved by Emerson in corn, by Blakeslee in *Portulaca,* and I have shown there is strong evidence for it in *Drosophila.* Hence these mutations are not caused by some general pervasive influence, but are due to "accidents" occurring on a molecular scale. When the molecular or atomic motions chance to take a particular form, to which the gene is vulnerable, then the mutation occurs.

It will even be possible to determine whether the entire gene changes at once, or whether the gene consists of several molecules or particles, one of which may change at a time. This point can be settled in organisms having determinate cleavage, by studies of the distribution of the mutant character in somatically mosaic mutants. If there is a group of particles in the gene, then when one particle changes it will be distributed irregularly among the descendant cells, owing to the random orientation of the two halves of the chromosome on the mitotic spindles of succeeding divisions,[5] but if there is only one particle to change, its mutation must affect all of the cells in a bloc, that are descended from the mutant cell.

(c) *The Conditions under which the Change occurs*

But the method that appears to have most scope and promise is the experimental one of investigating the conditions under which mutations occur. This requires studies of mutation frequency under various methods of handling the organisms. As yet, extremely little has been done along this line. That is because, in the past, a mutation was considered a windfall, and the expression "mutation frequency" would have seemed a contradiction in terms. To attempt to study it would have seemed as absurd as to study the conditions affecting the distribution of dollar bills on the sidewalk. You were simply fortunate if you found one. Not even controls, giving the "normal" rate of mutation—if indeed there is such a thing—were attempted.[6] Of late, however, we may say that certain very exceptional banking houses have been found, in front of which the dollars fall more frequently—in other words, especially mutable genes have been discovered, that are beginning to yield abundant data at the hands of Nilsson-Ehle, Zeleny,

Emerson, Anderson and others. For some of these mutable genes the rate of change is found to be so rapid that at the end of a few decades half of the genes descended from those originally present would have become changed. After these genes have once mutated, however, their previous mutability no longer holds. In addition to this "banking house method" there are also methods, employed by Altenburg and myself, for—as it were—automatically sweeping up wide areas of the streets and sifting the collections for the valuables. By these special genetic methods of reaping mutations we have recently shown that the ordinary genes of *Drosophila*—unlike the mutable genes above—would usually require at least a thousand years—probably very much more—before half of them became changed. This puts their stability about on a par with, if not much higher than, that of atoms of radium—to use a fairly familiar analogy. Since, even in these latter experiments, many of the mutations probably occurred within a relatively few rather highly mutable genes, it is likely that most of the genes have a stability far higher than this result suggests.

The above mutation rates are mere first gleanings—we have yet to find how different conditions affect the occurrence of mutations. There had so far been only the negative findings that mutation is not confined to one sex (Muller and Altenburg, 1919; Zeleny, 1921), or to any one stage in the life cycle (Bridges, 1919; Muller, 1920; Zeleny, 1921), Zeleny's finding that bar-mutation is not influenced by recency of origin of the gene (1921), and the as yet inconclusive differences found by Altenburg and myself for mutation rate at different temperatures (1919), until at this year's meeting of the botanists Emerson announced the definite discovery of the influence of a genetic factor in corn upon the mutation rate in its allelomorph, and Anderson the finding of an influence upon mutation in this same gene, caused by developmental conditions—the mutations from white to red of the mutable gene studied occurring far more frequently in the cells of the more mature ear than in those of the younger ear. These two results at least tell us decisively that mutation is not a sacred, inviolable, unapproachable process: it may be altered. These are the first steps; the way now lies open broad for exploration.

It is true that I have left out of account here the reported findings by several investigators, of genetic variations caused by treatments with various toxic substances and with certain other unusual conditions. In most of these cases, however, the claim has not been made that actual gene changes have been caused: the results have usually not been analyzed genetically and were in fact not analyzable genetically; they could just as well be interpreted to be due to abnormalities in the distribution of genes—for instance, chromosome abnormalities like those which Mavor has recently produced with X-rays—as to be due to actual gene mutations. But even if they were due to real genic differences, the possibility has in most cases by no means been excluded (1) that these genic differences were present in the stock to begin with, and merely became sorted out unequally, through random segregation; or (2) that other, invisible genic differences were present which, after random sorting out, themselves caused differences in mutation rate between the different lines. Certain recent results by Altenburg and myself suggest that genic differences, affecting mutation rate, may be not uncommon. To guard against either of these possibilities it would have been necessary to test the stocks out by a thorough course of inbreeding beforehand, or else to have run at least half a dozen different pairs of parallel lines of the control and treated series, and to have obtained a definite difference in the same direction between the two lines of *each* pair; otherwise it can be proved by the theory of "probable error" that the differences observed may have been a mere matter of random sampling among genic differences originally present. Accumulating large numbers of abnormal or inferior individuals by selective propagation of one or two of the treated lines—as has been done in some cases—adds nothing to the significance of the results.

At best, however, these genetically unrefined methods would be quite insensitive to mutations occurring at anything like ordinary frequency, or to such differences in mutation rate as have already been found in the analytical experiments on mutation frequency. And it seems quite possible that larger differences than these will not easily be hit upon, at least not in the early stages of our investigations, in view of the evidence that mutation is ordinarily due to an accident on an ultra-microscopic scale, rather than directly caused by influences pervading the organism. For the present, then, it appears most promising to employ organisms in which the genetic composition can be controlled and analyzed, and to use genetic methods that are sensitive enough to disclose mutations occurring in the control as well as in the treated individuals. In this way relatively slight variations in mutation frequency, caused by the special treatments, can be determined, and from the conditions found to alter the mutation rate slightly we might finally work up to those which affect it most markedly. The only methods now meeting this requirement are those in which a particular mutable gene is followed, and those in which many homozygous or else genetically controlled lines can be run in parallel, either by parthenogenesis, self-fertilization, balanced lethals or other special genetic means, and later analyzed through sexual reproduction, segregation and crossing over.

V. OTHER POSSIBILITIES

We cannot, however, set fixed limits to the possibilities of research. We should not wish to deny that some new and unusual method may at any time be found of directly producing mutations. For example, the phenomena now being worked out by Guyer may be a case in point. There is a curious analogy between the reactions of immunity and the phenomena of heredity, in apparently fundamental respects,[7] and any results that seem to connect the two are worth following to the limit.

Finally, there is a phenomenon related to immunity, of still more striking nature, which must not be neglected by geneticists. This is the d'Hérelle phenomenon. D'Hérelle found in 1917 that the presence of dysentery bacilli in the body caused the production there of a filterable substance, emitted in the stools, which had a lethal and in fact dissolving action on the corresponding type of bacteria, if a drop of it were applied to a colony of the bacteria that were under cultivation. So far, there would be nothing to distinguish this phenomenon from immunity. But he further found that when a drop of the affected colony was applied to a second living colony, the second colony would be killed; a drop from the second would kill a third colony, and so on indefinitely. In other words, the substance, when applied to colonies of bacteria, became multiplied or increased, and could be so increased indefinitely; it was self-propagable. It fulfills, then, the definition of an autocatalytic substance, and although it may really be of very different composition and work by a totally different mechanism from the genes in the chromosomes, it also fulfills our definition of a gene.[8] But the resemblance goes further —it has been found by Gratia that the substance may, through

appropriate treatments on other bacteria, become changed (so as to produce a somewhat different effect than before, and attack different bacteria) and still retain its self-propagable nature.

That two distinct kinds of substances—the d'Hérelle substances and the genes—should both possess this most remarkable property of heritable variation or "mutability," each working by a totally different mechanism, is quite conceivable, considering the complexity of protoplasm, yet it would seem a curious coincidence indeed. It would open up the possibility of two totally different kinds of life, working by different mechanisms. On the other hand, if these d'Hérelle bodies were really genes, fundamentally like our chromosome genes, they would give us an utterly new angle from which to attack the gene problem. They are filterable, to some extent isolable, can be handled in test tubes, and their properties, as shown by their effects on the bacteria, can then be studied after treatment. It would be very rash to call these bodies genes, and yet at present we must confess that there is no distinction known between the genes and them. Hence we cannot categorically deny that perhaps we may be able to grind genes in a mortar and cook them in a beaker after all. Must we geneticists become bacteriologists, physiological chemists and physicists, simultaneously with being zoologists and botantists? Let us hope so.

I have purposely tried to paint things in the rosiest possible colors. Actually, the work on the individual gene, and its mutation, is beset with tremendous difficulty. Such progress in it as has been made has been by minute steps and at the cost of infinite labor. Where results are thus meager, all thinking becomes almost equivalent to speculation. But we cannot give up thinking on that account, and thereby give up the intellectual incentive to our work. In fact, a wide, unhampered treatment of all possibilities is, in such cases, all the more imperative, in order that we may direct these labors of ours where they have most chance to count. We must provide eyes for action.

The real trouble comes when speculation masquerades as empirical fact. For those who cry out most loudly against "theories" and "hypotheses"—whether these latter be the chromosome theory, the factorial "hypothesis," the theory of crossing over, or any other—are often the very ones most guilty of stating their results in terms that make illegitimate *implicit* assumptions, which they themselves are scarcely aware of simply because they are opposed to dragging "speculation" into the open. Thus they

may be finally led into the worst blunders of all. Let us, then, frankly admit the uncertainty of many of the possibilities we have dealt with, using them as a spur to the real work.

1922

LITERATURE CITED

Blakeslee, A. F. 1920. *Genetics, 5:* 419–433.

Bridges, C. B. 1917. *Genetics, 2:* 445–465. 1919. *Proc. Soc. Exp. Biol. and Med., 17:* 1–2. 1921. *Proc. Nat. Acad. Sci. 7:* 186–192.

D'Hérelle, F. 1917. *Compt. rend. Acad., 165:* 373. 1918. *Compt. rend. Acad., 167:* 970. 1918. *Compt. rend. Soc. Biol., 81:* 1160. 1919. *Compt. rend. Acad., 168:* 631. 1920. *Compt. rend. Soc. Biol., 83:* 52, 97, 247.

Emerson, R. A. 1911. *Amer. Nat., 48:* 87–115.

Gratia, A., 1921. *Jour. Exp. Med., 34:* 115–126.

Mavor, J. W. 1921. *Science, N. S., 54:* 277–279.

Mohr, O. L. 1919. *Genetics, 4:* 275–282.

Muller, H. J. 1920. *Jour. Exp. Zool., 31:* 443–473.

Muller, H. J., and E. Altenburg. 1919. *Proc. Soc. Exp. Biol. and Med., 17:* 10–14. 1921. *Anat. Rec., 20:* 213.

Nilsson-Ehle, H. 1911. *Zeit. f. Ind. Abst. u. Vererb., 5:* 1–37. 1920. *Hereditas, 1:* 277–312.

Troland, L. T. 1917. *Amer. Nat., 51:* 321–350.

Wollstein, M. 1921. *Jour. Exp. Med., 34:* 467–477.

Zeleny, C. 1920. *Anat. Rec., 20:* 210. 1921. *Jour. Exp. Zool., 34:* 203–233.

NOTES

1 In symposium on "The Origin of Variations" at the thirty-ninth meeting of the American Society of Naturalists, Toronto, December 29, 1921.

2 It is of course conceivable, and even unavoidable, that *some* types of changes do destroy the gene's autocatalytic power, and thus result in its eventual loss.

3 It can hardly be true, as Troland intimates, that all similar fields attract each other more than they do dissimilar fields, otherwise all substances would be autocatalytic, and, in fact, no substances would be soluble. Moreover, if the parts of a molecule are in any kind of "solid," three-dimensional formation, it would seem that those in the middle would scarcely have opportunity to exert the moulding effect above mentioned. It therefore appears that a special manner of construction must be necessary, in order that a complicated structure like a gene may exert such an effect.

4 It may be that there are still other elements in the cell which have the

nature of genes, but as no critical evidence has ever been adduced for their existence, it would be highly hazardous to postulate them.

5 This depends on the assumption that if the gene does consist of several particles, the halves of the chromosomes, at each division, receive a random sample of these particles. That is almost a necessary assumption, since a gene formed of particles each one of which was separately partitioned at division would tend not to persist as such, for the occurrence of mutation in one particle after the other would in time differentiate the gene into a number of different genes consisting of one particle each.

6 Studies of "mutation frequency" had of course been made in the *Oenotheras,* but as we now know that these were not studies of the rate of gene change but of the frequencies of crossing over and of chromosome aberrations they may be neglected for our present purposes.

7 I refer here to the remarkable specificity with which a particular complex antigen calls forth processes that construct for it an antibody that is attracted to it and fits it "like lock and key," followed by further processes that cause more and more of the antibody to be reproduced. *If* the antigen were a gene, which could be slightly altered by the cell to form the antibody that neutralized it—as some enzymes can be slightly changed by heating so that they counteract the previous active enzyme—and if this antibody-gene then became implanted in the cell so as to keep on growing, all the phenomena of immunity would be produced.

8 D'Hérelle himself thought that the substance was a filterable virus parasitic on the bacterium, called forth by the host body. It has since been found that various bacteria each cause the production of D'Hérelle substances which are to some extent specific for the respective bacteria.

G. H. Hardy

Mendelian Proportions in a Mixed Population

The influence of Darwinism on British science was enormous. Detailed attention of the "continuous variations" in heredity led to a refined statistical analysis of populations. Francis Galton was the most prominent thinker in this field and a flourishing school of *biometry* was founded by him. The biometric outlook rejected Mendelism at first because its chief spokesman, Bateson, emphasized the discontinuity of heritable traits. Many years of controversy elapsed during the early period of the rediscovery of Mendelism before the biometric school and the Mendelian school found that both of their outlooks could be incorporated into a general theory of neo-Darwinism. G. H. Hardy (and C. Weinberg, independently, in Germany) pointed out the relevancy of Mendelism in biometric studies of evolution. The field of population genetics traces its origins to the Hardy-Weinberg law. It is a fallacy to think that a new mutation can swamp out a preexisting gene from the population in the absence of selection or in the absence of abnormal conditions. No matter how large the population becomes, the ratio of the mutant and normal genes in the population will remain constant. The study of the conditions under which the Hardy-Weinberg law fails to hold has been important in the study of evolution. With this law, population geneticists have worked out the frequency of genes for different ethnic groups in man. Anthropology relies heavily on this approach, using the human blood groups as an objective character since they can only be detected by immunological means.

Today the relation of genes in populations is expressed by the ratio $a^2 + 2ab + b^2 = 1$ where a^2 represents the frequency of individuals homozygous for *a* (a/a); 2ab represents the frequency of heterozygotes (a/b) in the population; and b^2 the frequency of homozygotes manifesting the *b* gene trait (b/b). If, for example nine percent of a human population is Rh negative, $b^2 = 9\%$ or $b = 30\%$ or 0.3 and since $a + b = 1$, $a = 0.7$ or 70% then

Reprinted from *Science 28*:49–50, 1908, by permission of the publisher.

$a^2 = 49\%$, $2ab = 42\%$ and the Hardy-Weinberg law presents the following information:

Frequency of gene *a* in population			0.7
"	"	*b* in population	0.3
"	"	a/a homozygotes in population	0.49
"	"	heterozygotes in population	0.42
"	"	b/b homozygotes in population	0.09

To the Editor of Science: I am reluctant to intrude in a discussion concerning matters of which I have no expert knowledge, and I should have expected the very simple point which I wish to make to have been familiar to biologists. However, some remarks of Mr. Udny Yule, to which Mr. R. C. Punnett has called my attention, suggest that it may still be worth making.

In the *Proceedings of the Royal Society of Medicine* (Vol. I., p. 165) Mr. Yule is reported to have suggested, as a criticism of the Mendelian position, that if brachydactyly is dominant "in the course of time one would expect, in the absence of counteracting factors, to get three brachydactylous persons to one normal."

It is not difficult to prove, however, that such an expectation would be quite groundless. Suppose that Aa is a pair of Mendelian characters, A being dominant, and that in any given generation the numbers of pure dominants (AA), heterozygotes (Aa), and pure recessives (aa) are as $p{:}2q{:}r$. Finally, suppose that the numbers are fairly large, so that the mating may be regarded as random, that the sexes are evenly distributed among the three varieties, and that all are equally fertile. A little mathematics of the multiplication-table type is enough to show that in the next generation the numbers will be as

$$(p + q)^2{:}2(p + q)(q + r){:}(q + r)^2, \text{ or as } p_1{:}2q_1{:}r_1, \text{ say.}$$

The interesting question is—in what circumstances will this distribution be the same as that in the generation before? It is easy to see that the condition for this is $q^2 = pr$. And since $q_1^2 = p_1r_1$, whatever the values of p, q and r may be, the distribution will in any case continue unchanged after the second generation.

Suppose, to take a definite instance, that A is brachydactyly, and that we start from a population of pure brachydactylous and pure normal persons, say in the ratio of 1:10,000. Then $p = 1$, $q = 0$, $r = 10{,}000$ and $p_1 = 1$, $q_1 = 10{,}000$, $r_1 = 100{,}000{,}000$. If

brachydactyly is dominant, the proportion of brachydactylous persons in the second generation is 20,001:100,020,001, or practically 2:10,000, twice that in the first generation; and this proportion will afterwards have no tendency whatever to increase. If, on the other hand, brachydactyly were recessive, the proportion in the second generation would be 1:100,020,001, or practically 1:100,000,000, and this proportion would afterwards have no tendency to decrease.

In a word, there is not the slightest foundation for the idea that a dominant character should show a tendency to spread over a whole population or that a recessive should tend to die out.

I ought perhaps to add a few words on the effect of the small deviations from the theoretical proportions which will, of course, occur in every generation. Such a distribution as $p_1:2q_1:r_1$, which satisfies the condition $q_1^2 = p_1r_1$, we may call a *stable* distribution. In actual fact we shall obtain in the second generation not $p_1:2q_1:r_1$ but a slightly different distribution $p_1':2q_1':r_1'$, which is not "stable." This should, according to theory, give us in the third generation a "stable" distribution $p_2:2q_2:r_2$, also differing slightly from $p_1:2q_1:r_1$; and so on. The sense in which the distribution $p_1:2q_1:r_1$ is "stable" is this, that if we allow for the effect of casual deviations in any subsequent generation, we should, according to theory, obtain at the next generation a new "stable" distribution differing but slightly from the original distribution.

I have, of course, considered only the very simplest hypotheses possible. Hypotheses other than that of purely random mating will give different results, and, of course, if, as appears to be the case sometimes, the character is not independent of that of sex, or has an influence on fertility, the whole question may be greatly complicated. But such complications seem to be irrelevant to the simple issue raised by Mr. Yule's remarks.

G. H. Hardy

Trinity College, Cambridge,
April 5, 1908

P.S. I understand from Mr. Punnett that he has submitted the substance of what I have said above to Mr. Yule, and that the latter would accept it as a satisfactory answer to the difficulty that he raised. The "stability" of the particular ratio 1:2:1 is recognized by Professor Karl Pearson (*Phil. Trans. Roy. Soc.* (A), vol. 203, p. 60).

1908

A. C. Allison

Protection Afforded by Sickle-Cell Trait Against Subtertian Malarial Infection

Sickle-cell anemia was shown to be a genetic defect by Neel; Pauling had come to that conclusion independently from his molecular analysis of hemoglobin obtained from adults showing the sickle-cell trait. But Allison noticed a peculiarity in the incidence of sickle-cell anemia. It was present in tropical areas of the world. It was a homozygous lethal condition and thus should be eliminated by natural selection. Why then did the disease persist?

Allison noted a similarity of the sickling-trait distribution and the incidence of malaria. He inferred that heterozygous carriers of the sickling trait had a resistance to malaria. Both of the homozygotes in the population were disadvantaged, one by lethality of the sickle-cell anemia, the other by susceptibility to malaria. In the Hardy-Weinberg law, the relation $a^2 + 2ab + b^2 = 1$ expresses a population equilibrium for the genes *a* and *b*. If *a* represents the normal hemoglobin and *b* the sickle-cell hemoglobin, then the selection against homozygous b *alone* would result in its disappearance. This is true, for example, among American descendants of African slaves who came from high malarial regions. Malaria will select against homozygous *a* but if the heterozygote, *ab*, is resistant to malaria, then the population equilibrium will favor heterozygous survival. Heterozygotes, however, cannot breed true to their genotype, and their progeny will include normal (a) and sickle-cell anemic (b) individuals. This condition is called "balanced polymorphism." It provides an example of human evolution because the removal of the malarial hazard predicts a subsequent decrease of the *b* gene in the population, as has already taken place in the United States.

Reprinted from *British Medical Journal 1*:290–294, 1954, by permission of the publisher and the author.

The aetiology of sickle-cell anaemia presents an outstanding problem common to both genetics and medicine. It is now universally accepted that the sickle-cell anomaly is caused by a single mutant gene which is responsible for the production of a type of haemoglobin differing in several important respects from normal adult haemoglobin (Pauling *et al.,* 1949; Perutz and Mitchison, 1950). Carriers of the sickle-cell trait who are heterozygous for the sickle-cell gene have a mixture of this relatively insoluble haemoglobin and normal haemoglobin; hence their erythrocytes do not sickle *in vivo,* whereas some at least of the homozygotes, who have a much greater proportion of sickle-cell haemoglobin, have sickle cells in the circulating blood, with inevitable haemolysis and a severe, often fatal, haemolytic anaemia. There is also a much smaller group of sickle-cell anaemia patients who are heterozygous for the sickle-cell gene as well as for some other hereditary abnormality of haemoglobin synthesis (Neel, 1952).

It is thus possible to approach the problem from the clinical or the genetical side. From the clinical point of view it is important to distinguish between carriers of the sickle-cell trait who show no other haematological abnormalities and patients with sickle-cell anaemia, who have a haemolytic disease which can reasonably be attributed to sickling of the erythrocytes. From the genetical point of view the main distinction is to be drawn between those who are homozygous and those who are heterozygous for the sickle-cell gene. In the great majority of instances two classifications coincide—that is, most individuals with the sickle-cell trait are heterozygous and most cases of sickle-cell anaemia, in Africa at least, are homozygous for the sickle-cell gene.

The sickle-cell trait is remarkably common in some parts of the world. Among many African Negro tribes 20% or more of the total population have the trait and frequencies of 40% have been found in several African tribes (Lehmann and Raper, 1949; Allison, 1954). In parts of Greece frequencies of 17% have been described (Choremis *et al.,* 1953), and as many as 30% of the population in Indian aboriginal groups are affected (Lehmann and Cutbush, 1952).

Wherever the sickle-cell trait is known to occur sickle-cell anaemia will also be found. For a time it was thought by some workers that sickle-cell anaemia was rare among African Negroes, but so many cases have been described during the past few years

that this view is no longer tenable (Lambotte-Legrand and Lambotte-Legrand, 1951; Foy *et al.,* 1951; Edington, 1953; Vandepitte and Louis, 1953).

The main problem can be stated briefly: how can the sickle-cell gene be maintained at such a high frequency among so many peoples in spite of the constant elimination of these genes through deaths from the anaemia? Since most sickle-cell anaemia subjects are homozygotes, the failure of each one to reproduce usually means the loss of two sickle-cell genes in every generation. It can be estimated that for the lost genes to be replaced by recurrent mutation so as to leave a balanced state, assuming that the sickle-cell trait—that is, the heterozygous condition—is neutral from the point of view of natural selection, it would be necessary to have a mutation rate of the order of 10^{-1}. This is about 3,000 times greater than naturally occurring mutation rates calculated for man and, with rare exceptions, in many other animals—3.2×10^{-5} in the case of haemophilia (Haldane, 1947). A mutation rate of this order of magnitude can reasonably be excluded as an explanation of the remarkably high frequencies of the sickle-cell trait observed in Africa and elsewhere.

POSSIBILITY OF SELECTIVE ADVANTAGE

Of the other explanations which can be advanced to meet the situation, one has received little attention: the possibility that individuals with the sickle-cell trait might under certain conditions have a selective advantage over those without the trait. It was stated for many years that the sickle-cell trait was in itself a cause of morbidity, but this belief seems to have been based upon unsatisfactory criteria for distinguishing the trait from sickle-cell anaemia. The current view is that the sickle-cell trait is devoid of selective value. Henderson and Thornell (1946) found that in American Negro air cadets who had passed a searching physical examination the incidence of the sickle-cell trait was the same as in the general Negro population of the United States. Lehmann and Milne (1949) were unable to discover any correlation between haemoglobin levels and the presence or absence of the sickle-cell trait in Uganda Africans. And Humphreys (1952) could find no evidence that the sickle-cell trait was responsible for any morbidity in Nigerian soldiers.

However, during the course of field work undertaken in Africa

in 1949 I was led to question the view that the sickle-cell trait is neutral from the point of view of natural selection and to reconsider the possibility that it is associated with a selective advantage. I noted then that the incidence of the sickle-cell trait was higher in regions where malaria was prevalent than elsewhere. The figures presented by Lehmann and Raper (1949) for the frequency of the sickle-cell trait in different parts of Uganda lent some support to this view, as did the published reports from elsewhere. Thus the trait is fairly common in parts of Italy and Greece, but rare in other European countries; in Greece the trait attains its highest frequencies in areas which are conspicuously malarious (Choremis *et al.*, 1951).

RELATION BETWEEN MALARIA AND SICKLE-CELL TRAIT

Other reports appeared suggesting more directly that there might be a relationship between malaria and the sickle-cell trait. Beet (1946) had observed that in a group of 102 sicklers from the Balovale district of Northern Rhodesia only 10 (9.8%) had blood slides showing malaria parasites, whereas in a comparable group of 491 non-sicklers 75 (15.3%) had malaria parasites. The difference in incidence of malaria in the two groups is statistically highly significant ($X^2 = 19.349$ for 1 d.f.)*; hence Beet's figures imply strongly that malaria is less frequent among individuals with the sickle-cell trait than among those without the trait. The difference in malarial susceptibility between sicklers and others seemed to be most pronounced at the time of the year when malaria transmission was lowest.

Later, in the Fort Jameson district of Northern Rhodesia, Beet (1947) found that the same difference was present, although it was much less pronounced. Of 1,019 non-sicklers, 312 (30.6%) had blood slides with malaria parasites, whereas of 149 sicklers 42 (28.2%) showed malaria parasites. This difference is not statistically significant. However, among the sicklers from Fort Jameson enlarged spleens were less common than among non-sicklers. In a series of 569 individuals there were 87 with the sickle-cell trait; 24 (27.9%) of these had palpable spleens, as compared with 188 (39.0%) with splenomegaly out of 482 non-sicklers. This difference is again statistically significant ($X^2 = 4.11$ for

* These and other statistics in this paper are my own, using available figures.

1 d.f.). Beet concluded that Africans with the sickle-cell trait were probably liable to recurrent attacks of thrombosis, with resultant shrinkage of the spleen.

Brain (1952a), also working in Rhodesia, confirmed Beet's observation that the spleen is palpable in a much lower proportion of sicklers than of non-sicklers; he went on to suggest that the finding might be explained by diminished susceptibility to malaria on the part of the sicklers. Moreover, Brain (1952b) compared the proportion of hospitalized cases in groups of African mine-workers with and without the sickle-cell trait. He found that the sicklers actually spent less time in hospital, on an average, than did the control group of non-sicklers. The incidence of malaria and pyrexias of unknown origin was much lower in the group with sickle cells.

It became imperative, then, to ascertain by more direct methods of investigation whether sickle cells can afford some degree of protection against malarial infection, thereby conferring a selective advantage on possessors of the sickle-cell trait in regions where malaria is hyperendemic. An opportunity to do this came during the course of a visit to East Africa in 1953.

INCIDENCE OF MALARIAL PARASITAEMIA IN AFRICAN CHILDREN WITH AND WITHOUT THE SICKLE-CELL TRAIT

The observations of Beet and of Brain on differences in parasite rates and spleen rates are open to criticism because the populations were heterogenous and were drawn from relatively wide areas. It was decided, therefore, to carry out similar tests on a relatively small circumscribed community, where all those under observation belong to a single tribe. Children were chosen rather than adults as subjects for the observations so as to minimize the effect of acquired immunity to malaria. The recorded incidence of parasitaemia in a group of 290 Ganda children, aged 5 months to 5 years, from the area surrounding Kampala (excluding the non-malarious township) is presented in Table I. The presence of sickling was demonstrated by chemical reduction of blood with isotonic sodium metabisulphite (Daland and Castle, 1948). Fresh reducing solutions were made up daily.

It is apparent that the incidence of parasitaemia is lower in the sickle-cell group than in the group without sickle-cells. The

TABLE I

	With Parasitaemia	Without Parasitaemia	Total
Sicklers	12 (27.9%)	31 (72.1%)	43
Non-sicklers	113 (45.7%)	134 (53.3%)	247

difference is statistically significant ($X^2 = 5.1$ for 1 d.f.). In order to test as many families as possible only one child was taken from each family. There is no reason to suppose that these groups are not comparable, apart from the presence or absence of the sickle-cell trait.

The parasite density in the two groups also differed: of 12 sicklers with malaria, 8 (66.7%) had only slight parasitaemia (group 1 on an arbitrary rating), while 4 (33.3%) had a moderate parasitaemia (group 2). Of the 113 non-sicklers with malaria, 34% had slight parasitaemia (group 1), the parasite density in the remainder being moderate or severe (group 2 or 3).

It may be noted, incidentally, that of the four cases in the sickle-cell group with moderate parasitaemia three had *P. malariae,* even though this species is much less common than *P. falciparum* around Kampala. It seems possible from these and other observations that the protection afforded by the sickle-cell trait is more effective against *P. falciparum* than against other species of plasmodia, but much further work is necessary to decide the point.

These results suggest that African children with the sickle-cell trait have malaria less frequently or for shorter periods, and perhaps also less severely, than children without the trait. Further evidence regarding the protective action of the sickle-cell trait could be obtained only by direct observation on the course of artificially induced malarial infection in volunteers.

PROGRESS OF MALARIAL INFECTION IN ADULT AFRICANS WITH AND WITHOUT THE SICKLE-CELL TRAIT

Fifteen Luo with the trait and 15 Luo without the trait were accepted for this investigation. All of the volunteers were adult males who had been away from a malarious environment for

TABLE II

No.	Mode of Infection and Strain	Day after Infection 8	10	12	14	16	18	20	22	24	26	28	30	32	34	36	38	40	
									Luo with No Sickle Cells										
1	M_2 B_1	0.03	—	0.07	2.5	5.0	2.5	5.0	1.2	0.4	0.02	0.01	—	—	0.1	0.01	0.01	ST	
2	M_2 B_1	—	—	—	—	—	—	—	0.03	0.13	0.41	—	—	—	—	0.03	—	ST	
3	M_2 B_1	—	—	—	—	—	—	—	0.1	0.02	0.20	5.0	2.5	1.25	1.67	0.2	5.0	2.0	ST
4	M_2 B_1	—	—	—	—	0.02	0.02	0.5	0.83	0.12	0.2	1.0	1.0	0.83	0.25	0.17	—	ST	
5	M_2 B_2	—	—	—	—	0.05	1.0	1.67	0.25	0.05	0.07	0.25	1.2	1.0	0.03	—	—	ST	
6	B_1	0.02	5.0	10.0	10.0	1.0	0.1	0.01	ST	—	—	—	—	—	—	—	—	—	
7	B_2	—	—	—	—	15.0	50.0	ST	—	—	—	—	—	—	—	—	—	—	
8	B_2	—	—	0.13	5.0	1.67	0.33	—	—	ST	—	—	—	—	—	—	—	—	
9	B_1	—	—	—	—	5.0	—	0.1	0.5	2.5	—	1.0	0.1	2.5	10.0	5.0	0.5	ST	
10	B_2	—	—	—	—	—	—	0.05	0.05	—	—	0.67	—	0.1	0.05	5.0	5.0	ST	
11	B_2	—	0.05	0.3	—	—	—	0.2	ST										
12	B_2	—	—	0.3	0.3	0.3	0.1	0.3	ST										
13	B_2	—	—	—	—	—	—	—	—	—	—	—	—	—	—	—	ST		
14	B_2	2.0	1.7	2.0	60.0	5.0	0.6	ST											
15	B_2	0.05	0.3	—	0.4	0.1	0.3	ST											
									Luo with Sickle-Cell Trait										
1	M_1 B_2	—	—	—	—	—	—	—	—	—	—	—	—	—	—	—	—	ST	
2	M_1 B_2	—	—	—	—	—	—	—	—	—	—	—	—	—	—	—	—	″	
3	M_1 B_1	—	—	—	—	—	—	—	—	—	—	—	—	—	—	—	—	″	
4	M_1 B_1	—	—	—	—	—	—	—	—	—	—	—	—	—	—	—	—	″	
5	M_1 B_1	—	—	—	—	—	—	—	—	—	—	—	—	—	—	—	—	″	
6	M_1 B_1	—	—	—	—	—	—	—	—	—	—	—	—	—	—	—	—	″	
7	M_1 B_1	—	—	—	—	—	—	—	—	—	—	—	—	—	—	5.0	0.5	″	
8	M_1 B_1	—	—	—	—	—	—	—	—	—	—	—	—	—	—	—	—	″	
9	M_1 B_1	0.7	—	—	—	—	—	—	—	—	—	0.03	0.1	0.03	0.03	—	—	″	
10	M_1 B_1	—	—	—	—	—	—	—	—	—	—	—	—	—	—	—	—	″	
11	B_2 M_2	—	—	—	—	—	—	—	—	—	—	—	—	—	—	—	—	″	
12	B_2 M_2	—	—	—	—	—	—	—	—	—	—	—	—	—	—	—	—	″	
13	B_2 M_2	—	—	—	—	—	—	—	—	—	—	—	—	—	—	—	—	″	
14	B_2 M_2	—	—	—	—	—	—	—	—	—	—	—	—	—	—	—	—	″	
15	B_2 M_2	—	—	—	—	—	—	—	—	—	—	—	—	—	—	—	—	″	

Figures represent parasite counts in hundreds per mm.^{3} of blood. ST=Stopped by chemotherapy.

at least 18 months. The two groups were of a similar age and appeared to be strictly comparable apart from the presence or absence of the sickle-cell trait. Two strains of *P. falciparum* were used—one originally isolated in Malaya and one from near Mombasa, Kenya; in Table II these are marked with the subscripts 1 and 2 respectively. The infection was introduced by sub-inoculation with 15 ml. of blood containing a large number of trophozoites (B in the table) or by biting with heavily infected *Anopheles gambiae* (M in the table). At least 3 out of the 10 mosquitoes applied had bitten each individual, and the presence of sporozoites was confirmed by dissection of the mosquitoes.

The cases were followed for 40 days. Parasite counts for each case were made by comparison with the number of leucocytes in 200 oil-immersion fields of thick films, the absolute leucocyte counts being checked at intervals. The abbreviated results of these counts are shown in Table II. In the few cases in which parasitaemia was pronounced and the symptoms were relatively severe the progress of the disease was arrested. At the end of the period of observation in every case a prolonged course of antimalarial chemotherapy was given.

DISCUSSION

It is apparent that the infection has become established in 14 cases without the sickle-cell trait. The parasitaemia is relatively light, however, when compared with that observed in non-immune populations—for example, the Africans described by Thomas *et al.* (1953). This is to be expected: the Luo come from a part of the country where malaria is hyperendemic, and they have acquired a considerable immunity to malarial infection in childhood. This factor makes the interpretation of the observations rather more difficult; however, it could not be avoided, since all the East African tribes who have high sickling rates come from malarious areas, and the acquired immunity should operate with equal force in the groups with and without sickle-cells. The acquired immunity was actually an advantage, since the symptoms were mild and the chances of complication very slight.

In the group with sickle cells, on the other hand, the malaria parasites have obviously had great difficulty in establishing them-

selves, in spite of repeated artificial infection. Only two of the cases show parasites, and the parasite counts in these cases are comparatively low. The striking difference in the progress of malarial infection in the two groups is taken as evidence that the abnormal erythrocytes in individuals with the sickle-cell trait are less easily parasitized than are normal erythrocytes.

It can therefore be concluded that individuals with the sickle-cell trait will, in all probability, suffer from malaria less often and less severely than those without the trait. Hence in areas where malaria is hyperendemic children having the trait will tend to survive, while some children without the trait are eliminated before they acquire a solid immunity to malarial infection. The protection against malaria might also increase the fertility of possessors of the trait. The proportion of individuals with sickle cells in any population, then, will be the result of a balance between two factors; the severity of malaria, which will tend to increase the frequency of the gene, and the rate of elimination of the sickle-cell genes in individuals dying of sickle-cell anaemia. Or, genetically speaking, this is a balanced polymorphism where the heterozygote has an advantage over either homozygote.

The incidence of the trait in East Africa has recently been investigated in detail (Allison, 1954), and found to vary in accordance with the above hypothesis. High frequencies are observed among the tribes living in regions where malaria is hyperendemic (for example, around Lake Victoria and in the Eastern Coastal Belt), whereas low frequencies occur consistently in the malaria-free or epidemic zones (for example, the Kigezi district of Uganda; the Kenya Highlands; and the Kilimanjaro, Mount Meru, and Usumbara regions of Tanganyika). This difference is often independent of ethnic and linguistic grouping: thus, the incidence of the sickle-cell trait among Bantu-speaking tribes ranges from 0 (among the Kamba, Chagga, etc.) to 40% (among the Amba, Simbiti, etc.). The world distribution of the sickle-cell trait is also in accordance with the view presented here that malarial endemicity is a very important factor in determining the frequency of the sickle-cell trait. The genetical and anthropological implications of this view are evident.

The fact that sickle cells should be less easily parasitized by plasmodia than are normal erythrocytes is presumably attributable to their haemoglobin component, although there may be other differences, not yet observed, between the two cell-types.

Sickle-cell haemoglobin is unlike normal adult haemoglobin in important physico-chemical properties, notably in the relative insolubility of the sickle-cell haemoglobin when reduced (Perutz and Mitchison, 1950). The malaria parasite is able to metabolize haemoglobin very completely in the intact red cell, the haematin pigment remaining as a by-product of haemoglobin breakdown (Fairley and Bromfield, 1934; Moulder and Evans, 1946). That plasmodia are greatly affected by relatively small differences in their environment is suggested by their remarkable species specificity. Thus the difficulty of establishing an infection in monkeys with human malaria parasites, and vice versa, is generally recognized.

How far species differences in the haemoglobins themselves, known from immunological and other studies, are responsible for the species specificity of plasmodia it is impossible to say. However, the physico-chemical differences between human adult haemoglobin and monkey haemoglobin appear to be less pronounced than the differences between either type and sickle-cell haemoglobin. It is clear that the natural resistance to malaria among individuals with the sickle-cell trait is relative, not absolute. This is perhaps attributable to differences in the expressivity of the sickle-cell gene, which may be responsible for the production of from nearly 50% to only a very small amount of sickle-cell haemoglobin (Wells and Itano, 1951; Singer and Fisher, 1953). Moreover, the sickle-cell haemoglobin may not be evenly distributed in the cell population: most observers recognize that there are cases in which only some of the red cells are sickled even after prolonged reduction. However, even a relative resistance to malaria may be enough to help those with the sickle-cell trait through the dangerous years of early childhood, during which an active immunity to the disease is developed.

The above observations focus attention upon the importance of haemoglobin to plasmodia in the erythrocytic phase. Hence it is worth considering whether erythrocytes containing other specialized or abnormal types of haemoglobin might be resistant to malaria also. Thus, human foetal haemoglobin differs from human adult haemoglobin in many properties. Red cells containing foetal haemoglobin continue to circulate in the newborn for the first three months of life, after which they are quite rapidly replaced by cells containing normal adult haemoglobin. It has long been known that the newborn has a considerable de-

gree of resistance to malarial infection: Garnham (1949), for instance, found that in the Kavirondo district of Kenya at the end of the second month of life only 10% of babies were infected; after this age the percentage affected rises rapidly, until by the ninth month practically all children have the disease. The correspondence between the appearance of cells containing normal adult-type haemoglobin and malarial susceptibility is illustrated in the chart. The correspondence may of course be fortuitous, but it is striking enough to merit further investigation, even though other factors, such as a milk diet deficient in *p*-aminobenzoic acid (Maegraith *et al.*, 1952; Hawking, 1953) and immunity acquired from the mother (Hackett, 1941) may play a part in the natural resistance of the newborn to malaria.

Finally, it is possible that the explanation offered above for the maintenance of the sickle-cell trait may also apply to thalassaemia. The problems presented by the two diseases are very similar; many homozygotes, and possibly some heterozygotes, are known to die of thalassaemia, and yet the condition remains remarkably common in Italy and Greece, where as many as 10% of the individuals in certain areas are affected (Bianco *et al.*, 1952). Greek and Italian authors have commented that cases of thalassaemia usually come from districts severely afflicted with malaria (Choremis *et al.*, 1951). Perhaps those who are heterozygous for the thalassaemia gene suffer less from malaria than their compatriots; the fertility of the heterozygotes appears to be greater (Bianco *et al.*, 1952). Selective advantage of the heterozygote with the sickle-cell gene, and possibly the heterozygote with the thalassaemia gene also, would explain why such high gene frequencies can be attained in the case of these conditions while other genetically transmitted abnormalities of the blood cells remain uncommon, not very much above the estimated mutation rate—for example, hereditary spherocytosis (Race, 1942).

SUMMARY

A study has been made of the relationship between the sickle-cell trait and subtertian malarial infection. It has been found that in indigenous East Africans the sickle-cell trait affords a considerable degree of protection against subtertian malaria. The incidence of parasitaemia in 43 Ganda children with the sickle-cell trait was significantly lower than in a comparable group of

247 children without the trait. An infection with *P. falciparum* was established in 14 out of 15 Africans without the sickle-cell trait, whereas in a comparable group of 15 Africans with the trait only 2 developed parasites.

It is concluded that the abnormal erythrocytes of individuals with the sickle-cell trait are less easily parasitized by *P. falciparum* than are normal erythrocytes. Hence those who are heterozygous for the sickle-cell gene will have a selective advantage in regions where malaria is hyperendemic. This fact may explain why the sickle-cell gene remains common in these areas in spite of the elimination of genes in patients dying of sickle-cell anaemia. The implications of these observations in other branches of haematology are discussed.

This investigation was made possible by a grant received from the Colonial Office at the recommendation of the Colonial Medical Research Committee. Acknowledgment is made to the Directors of Medical Services, Kenya and Uganda, to Dr. E. J. Foley, Nairobi, to the staff of the Mulago Hospital, Kampala, and to the volunteers for their kind co-operation. To Dr. G. I. Robertson, Nairobi, a special debt of gratitude is due for constant help and advice. Drs. A. E. Mourant, R. G. Macfarlane, and A. H. T. Robb-Smith read the manuscript and made valuable comments and suggestions.

1954

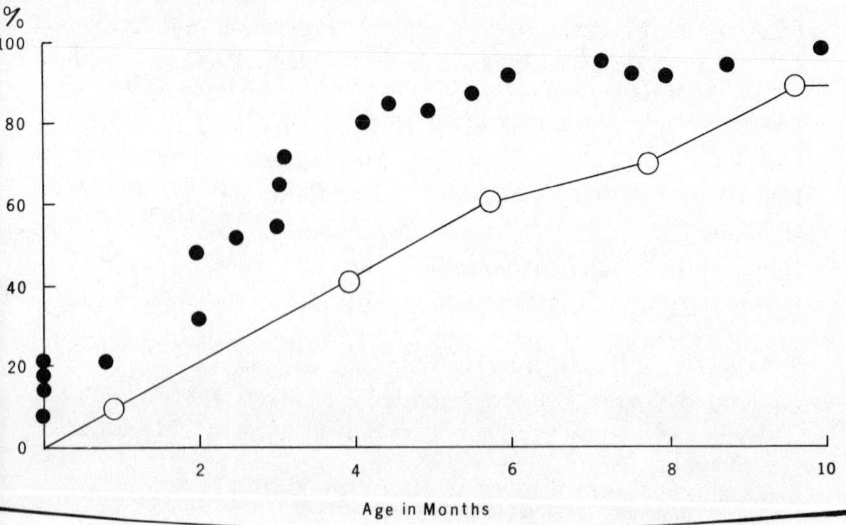

FIG. 1. The apparent relationship between the appearance of adult-type haemoglobin (dots) and malarial infection (circles) in the newborn. Each dot represents a test on a single individual, using an alkali denaturation technique (Allison, unpublished observations); the circles represent the percentage of Luo children showing malaria parasites (Garnham, 1949).

REFERENCES

Allison, A. C. (1954). *Trans. roy. Soc. trop. Med. Hyg.* In press.

Beet, E. A. (1946). *E. Afr. med. J.,* 23, 75.

——— (1947). *Ibid.,* 24, 212.

Bianco, I., Montalenti, G., Silvestroni, E., and Siniscalo, M. (1952). *Ann. Eugen. Lond.,* 16, 299.

Brain, P. (1952a). *British Medical Journal,* 2, 880.

———(1952b). *S. Afr. med. J.,* 26, 925.

Choremis, C., Zervos, N., Constantinides, V., and Zannos, Leda (1951). *Lancet,* 1, 1147.

Choremis, C., Ikin, Elizabeth W., Lehmann, H., Mourant, A. E., and Zannos, Leda (1953). *Ibid.,* 2, 909.

Daland, G. A., and Castle, W. B. (1948). *J. Lab. clin. Med.,* 33, 1082.

Edington, G. M. (1953). *British Medical Journal,* 2, 957.

Fairley, Sir N. H., and Bromfield, R. J. (1934). *Trans. roy. Soc. trop. Med. Hyg.,* 28, 141.

Foy, H., Kondi, A., and Brass, W. (1951). *E. Afr. med. J.,* 28, 1.

Graham, P. C. C. (1949). *Ann. trop. Med. Parasit.,* 43, 47.

Hackett, L. W. (1941). *Publ. Amer. Ass. Advanc. Sci.,* 15, 148.

Haldane, J. B. S. (1947). *Ann. Eugen., Lond.,* 13, 262.

Hawking, F. (1953). *British Medical Journal,* 1, 1201.

Henderson, A. B., and Thornell, H. E. (1946). *J. Lab. clin. Med.,* 31, 769.

Humphreys, J. (1952). *J. trop. Med. Hyg.,* 55, 166.

Lambotte-Legrand, J., and Lambotte-Legrand, C. (1951). *Institut Royal Colonial Belge, Sect. d. Sci. Nat. et Med.,* Memoires, Tome XIX, fasc. 7, p. 98.

Lehmann, H., and Cutbush, Marie (1952). *British Medical Journal,* 1, 404.

Lehmann, H., and Milne, A. H. (1949). *E. Afr. med. J.,* 26, 247.

Lehmann, H., and Raper, A. B. (1949). *Nature, Lond.,* 164, 494.

Maegraith, B. G., Deegan, T., and Jones, E. S. (1952). *British Medical Journal,* 2, 1382.

Moulder, J. W., and Evans, E. A. (1946). *J. biol. Chem.,* 164, 145.

Neel, J. V. (1950). *Cold Spr. Harb. Symp. quant. Biol.,* 15, 141.

———(1952). *Blood,* 7, 467.

Pauling, L., Itano, H. A., Singer, S. J., and Wells, I. C. (1949). *Science,* 110, 543.

Perutz, M. F. and Mitchison, J. M. (1950). *Nature, Lond.,* 166, 677.

Race, R. R. (1942). *Ann. Eugen. Lond.,* 11, 365.

Singer, K., and Fisher, B. (1953). *Blood,* 8, 270.

Thomas, A. T. G., Robertson, G. I., and Davey, D. G. (1953). *Trans. roy. Soc. trop. Med. Hyg.,* 47, 338.

Vandepitte, J. M., and Louis, L. A. (1953). *Lancet,* 2, 806.

Wells, I. C., and Itano, H. A. (1951). *J. biol. Chem.,* 188, 65.

E. Zuckerkandl and W. A. Schroeder

Amino-acid Composition of the Polypeptide Chains of Gorilla Haemoglobin

Although man is not a suitable organism for genetic research, much has been learned about his heredity through the biochemical analysis of his hemoglobin. Numerous hemoglobin abnormalities throughout the world have been analyzed and more than thirty different amino-acid replacements have been identified. Since hemoglobin is found in all mammals and many other forms of life, the techniques of Pauling and Ingram have been used widely in a comparative study of the hemoglobin molecule. This approach greatly increases the scientific analysis of evolution. If molecules evolve through mutation and selection, differences in these molecules should be reflected in man's ancestors. This paper by Zuckerkandl and Schroeder illustrates the relation of man's hemoglobin to his closest "relatives," the primates. In the future there will be many more studies of the evolution of proteins. It is a difficult, but rewarding, field which provides students of evolution theory with the technology of molecular biology.

Comparison of tryptic peptide patterns has demonstrated the great overall similarity of the primary structure of haemoglobins from normal adult man and anthropoid apes.[1] Work is now in progress to define this similarity more precisely. We report here the determination of the amino-acid composition of the polypeptide chains of gorilla haemoglobin.

Reprinted from *Nature 192*:984–985, 1961, by permission of the publisher and the authors.

Recent investigation in these laboratories has shown that gorilla haemoglobin contains two pairs of polypeptide chains. These have the same N-terminal sequence by the dinitrophenyl method as the polypeptide chains of human haemoglobin.[2, 3] We designate them, therefore, the α- and β-chains.

The α- and β-chains of gorilla haemoglobin were separated by chromatography according to the method of Wilson and Smith.[4] Individual chains were hydrolysed in 6 N hydrochloric acid at 110° C. for 22 and 70 hr. The amino-acid composition was determined by means of a Spinco automatic amino-acid analyser.

The results are given in Table I.

On the basis of the amino-acid composition, the α- and β-chains of gorilla haemoglobin are almost indistinguishable from the corresponding chains of human haemoglobin. The α-chain of gorilla haemoglobin appears to have one more residue of aspartic acid and one less residue each of glutamic acid and serine. Amino-acid analysis of tryptic peptides from the α-chain has revealed the substitution of an aspartyl for a glutamyl residue in a peptide equivalent to spot No. 23 (ref. 8) or peptide αT-4 (ref. 9) of the α-chain of human haemoglobin. Assuming an essentially identical sequence in the two types of α-chains, this substitution probably is in the 23rd residue from the N-terminus. Whether the α-chain of gorilla haemoglobin actually contains one less seryl residue is uncertain because of the destruction of serine during hydrolysis. The value given, however, was obtained from each of two independent determinations of the destruction of serine as a function of the time of hydrolysis.

The β-chain of gorilla haemoglobin clearly contains one less arginyl residue than that of human haemoglobin and may contain an additional lysyl residue.

In the hydrolysates of the β-chain, homocitrulline was found. Presumably, this was formed by reaction of lysine with cyanate[10] in the urea which was used in the separation of the chains although its formation would not have been anticipated because of the acidity of the solutions. The value for lysine in the β-chain thus is reported as the sum of lysine and homocitrulline.

These amino-acid compositions thus yield further evidence of the similarity of gorilla and human haemoglobins. The substitution of an aspartyl for a glutamyl residue in the α-chain and the possible substitution of a lysyl for an arginyl residue in the β-chain are compensatory in charge in each chain. As a result, the

TABLE I

NUMBER OF RESIDUES PER α- AND β-CHAIN OF GORILLA AND HUMAN HAEMOGLOBINS

	α-chain		β-chain	
Amino-acid	Human*	Gorilla	Human*	Gorilla
ala	21	21	15	15
arg	3	3	3	2
asp	12	13	13	13
cys/2	1	1	2	2
glu	5	4	11	11
gly	7	7	13	13
his	10	10	9	9
ileu	0	0	0	0
leu	18	18	18	18
lys	11	11	11	11 or 12
met	2	2	1	1
phe	7	7	8	8
pro	7	7	7	7
ser	11	10	5	5
thr	9	9	7	7
try	1	1†	2	2†
tyr	3	3	3	3
val	13	13	18	18
Total	141	140	146	145 or 146

* Based on refs. 5 and 6 for the α-chain, on ref. 7 for the β-chain, and on a personal communication from G. Braunitzer.

† Estimated from the tryptic peptide pattern and from the ultraviolet spectrum.

electrophoretic behaviours should be alike. Indeed, migration-rates of the two haemoglobins on paper electrophoresis at pH 8·6 are indistinguishable. (Paper electrophoretic migrations were kindly determined by Dr. W. R. Bergren.) The chromatographic behaviours on 'Amberlite IRC-50' are detectably, but very slightly, different.[11] Whether the above are the sole differences in composition is, of course, to some extent uncertain because of the usual inaccuracies of amino-acid analysis; for this reason, one cannot be sure that the serine content of the α-chains is different. The amino-acid composition in no way can provide information about changes in sequence without changes in composition.

While these restrictions must be kept in mind, the α-chains of gorilla and human haemoglobin may well differ by only two residues and the β-chains by one residue. The differences, then, appear to be of the order observed between normal and abnormal human haemoglobins. Indeed, abnormal human haemoglobins such as haemoglobin S or haemoglobin I differ from normal human haemoglobin apparently only by one amino-acid substitution in one or the other chain.[12, 13] One might say that the β-chain especially has the characteristics of an abnormal human β-chain although a human β-chain corresponding to the gorilla β-chain has not yet been found. The results suggest that the haemoglobin genes of man and gorilla constitute a single population.

We wish to express our appreciation of valuable discussions with Prof. Linus Pauling. Mrs. Joan Balog Shelton and Mr. Johnson Cua have given competent technical assistance.

This work was supported in part by a grant (H-2558) from the National Institutes of Health, U.S. Public Health Service. One of us (E.Z.) is on leave from the Centre National de la Recherche Scientifique, Paris.

1961

NOTES

[1] Zuckerkandl, E., Jones, R. T., and Pauling, L., *Proc. U.S. Nat. Acad. Sci.,* 46, 1349 (1960).

[2] Rhinesmith, H. S., Schroeder, W. A., and Pauling, L., *J. Amer. Chem. Soc.,* 79, 609 (1957).

[3] Rhinesmith, H. S., Schroeder, W. A., and Martin, N., *J. Amer. Chem. Soc.,* 80, 3358 (1958).

[4] Wilson, S., and Smith, D. B., *Canad. J. Biochem. Physiol.,* 37, 405 (1959).

[5] Braunitzer, G., Rudloff, V., Hilse, K., Liebold, B., and Müller, R., *Z. physiol. Chem.,* 320, 283 (1960).

[6] Konigsberg, W., Guidotti, G., and Hill, R. J., *J. Biol. Chem.,* 236, PC, 55 (1961).

[7] Braunitzer, G., Hilschmann, N., Hilse, K., Liebold, B., and Müller, R., *Z. physiol. Chem.,* 322, 96 (1960).

[8] Ingram, V. M., *Biochim. Biophys. Acta.,* 28, 539 (1958).

[9] Baglioni, C., *Biochim. Biophys. Acta,* 48, 392 (1961).

[10] Stark, G. R., Stein, W. H., and Moore, S., *J. Biol. Chem.,* 235, 3177 (1960).

[11] Jones, R. T., Ph.D. thesis, Calif. Inst. Tech. (1961).

[12] Ingram, V. M., *Biochim. Biophys. Acta,* 36, 402 (1959).

[13] Murayama, M., *Fed. Proc.,* 19, 78 (1960).

Glossary and Subject Index

adaptive enzyme: an enzyme which is not normally present in an organism but which greatly increases in amount when a new substrate (food) is present as its major organic source. 182

allele: a variant form of a gene, or a normal gene and its various alleles (in older literature, its synonym is *allelomorph*). 62, 69, 269

amino acid: the smallest functional component of a protein. There are twenty major amino acids all having the structure

$$H_2N\text{—}\overset{\displaystyle R}{\underset{\displaystyle H}{\overset{|}{\underset{|}{C}}}}\text{—}COOH$$

where R represents any one of twenty different types of attachment. 5, 73, 167, 168, 176, 199

balanced polymorphism: the presence of two alleles (or chromosomal collections of non-recombining genes) in heterozygous condition with the heterozygote more adapted to its environment than either of its homozygous forms. 305, 313

base analogue: a synthetic or unnatural variant of normally occurring purines and pyrimidines. The chemical properties of base analogues result in *transition* mutations. 207, 210, 212, 222

base pair: the complementary nitrogenous bases in DNA or RNA. Adenine and thymine form one base pair; guanine and cytosine form a second base pair. In RNA, adenine and uracil constitute a base pair. Hydrogen bonds establish the proper pairing. 84, 89, 195, 210, 212, 217

blastomere: an early embryonic cell formed by the first few mitoses or cleavages of a fertilized egg. 115

cell: the smallest unit of a living organism which can carry out metabolism. In most plants and animals the cell is a porous membranous system containing a nucleus within which genetic material is stored and utilized. 1, 2, 17, 23, 26, 29, 33, 39

cell doctrine: the view that all cells arise from pre-existing cells. The process accomplishing this is called *mitosis.* 3, 22, 23

cell fractionation: the process of degrading a cell into its components and using different purified fractions for biochemical and biophysical studies. 18

chromatin: the diffuse, lint-like appearance of unwound chromosomes in fixed and stained cells. 27

chromosome: a contiguous group of genes usually aligned linearly in higher plants and animals. The genic material of the chromosome of cells is DNA; in most cells the chromosome is tightly coiled during mitosis and gives it a compact shape. 4, 5, 7, 11, 13, 26, 37, 61, 99, 231, 286, 288, 291

cistron: the gene defined functionally, with recombination tests establishing an internal fine structure. The modern word for a *gene,* but in precise usage, the gene as a defined unit of function. 93, 145, 146, 183

codon: a sequence of three nucleotides on a chain of DNA or messenger RNA which encodes one amino acid. 150, 151, 222

colinearity: the theory which relates the linear sequence of nucleotides to a linear sequence of amino acids through a genetic coding mechanism. The colinearity also extends to messenger RNA and to the map of mutant sites within a gene. 8, 174, 192, 222

complementation: the expression of a normal phenotype by a heterozygote, whether the genes involved are allelic or nonallelic. 144

continuous variation (also called *fluctuating variation*): the imperceptible gradation of differences for quantitative traits such as human height or intelligence. Usually a consequence of genetic (multiple factor) and environmental (non-inherited) components. 233, 261, 283

convariant reproduction: the capacity of a gene and its mutant alleles to make copies of themselves, providing the paradox that mutational changes may drastically alter a gene's function but they do not affect the gene's capacity to copy its errors. The Watson–Crick theory provides the mechanism for this phenomenon. 3, 10, 84

coupling: the linkage or association of genes, now known to be a consequence of their organization into chromosomes. The coupled genes are separable by crossing over and then act as if they were in a state of repulsion. 61, 62

crossing over: genetic recombination in homologous chromosomes during meiosis resulting in the exchange of paternal and maternal segments of the homologues. 13, 61, 93

deoxyribonucleic acid (DNA): the chemical substance which stores genetic information; genic material. 2, 4, 7, 11, 12, 16, 38, 84, 85, 89, 143, 149, 150, 174, 182, 192, 193, 205, 211, 212, 213, 222

development: the biological process leading to complex changes in cellular or noncellular systems; in higher plants and animals—embryology. 5

differentiation: a process of intracellular change in protein composition leading to tissue formation. 99, 104, 114, 122, 128

diploid: the chromosome number formed by the fertilization of an egg by a sperm; diploid cells contain two haploid sets of chromosomes. Abbreviated, 2n, in man: 2n = 46. 5, 7, 9

endoplasmic reticulum: a network of membranes bounded by the cell surface and the nucleus; the membranes contain ribosomes which are active in protein synthesis. 2, 11, 184

epigenesis: the embryological theory that complexity increases from fertilization to birth by progressive changes in differentiation, growth, and morphogenetic movement. 5, 29, 30, 98

equipotential system: an embryonic organization in which separated blastomeres or experimentally segmented portions of the embryo give rise to complete embryos. 119, 121

eugenics: the application of selective genetic breeding to man by voluntary or involuntary means with the expectation that human evolution can be directed to human betterment. 65, 70, 72

evolution: a theory of complexity in the organization of life from the origins of life to the present with the premise that all life is related by common descent to the first forms of life on earth. 6, 229, 257, 260, 264, 267, 319

fine structure: the resolution, by recombination, of sites within a gene. The maximum fineness corresponds to dimensions of 1 or 2 nucleotides along the linear length of DNA. 93, 143, 205

fingerprints: the smudges on a chromatogram formed by different proteins or their fragments. 167, 168, 170

gene: the conceptual unit of heredity; structurally it is found in chromosomes;

its genetic storage material is DNA; it functions by generating its code through RNA copies which are translated into proteins by the endoplasmic reticulum or ribosomes. 3, 4, 14, 34, 37, 40, 61, 64, 65, 68, 71, 73, 75, 84, 93, 94, 138, 139, 141, 168, 171, 174, 183, 192, 238, 268, 285

genetic code: the process by which information in DNA nucleotide sequences is translated into protein amino acid sequences. 93, 150, 174, 191, 198, 200, 222, 223, 227

genetic map: the points or sites representing the gene or mutant site on a diagram corresponding to a chromosome. The map distances are determined by the frequency with which crossing over occurs between two sites. 15, 96

genome: the complete set of all the genes in an organism; its total stored genetic information.

genotype: the genetic constitution of an organism for one or more discernible hereditary traits. 138, 268

germ plasm: the gametes and their immature tissue states found in the testes or ovaries. 36, 99, 114, 118

haploid: the chromosome number of a gamete or a single set of chromosomes; abbreviated n; in man: $n = 23$. 5, 7, 9

histology: the study of tissues and the cellular composition of tissues. 23

homologous chromosomes: in a diploid cell the compact chromosomes can be paired physically or optically. Such pairs of chromosomes are homologues. They contain similar sequences of genes. 13, 62

hydrogen bond: a weak atomic bond between hydrogen protons and negatively charged (but not ionized) molecules. 86, 90, 91

idioplasm: a nineteenth century conceptual unit of heredity; later replaced by the term *genetic material.* 28, 31

independent assortment: the genetic segregation of related genes which are carried by non-homologous chromosomes; the direction of chromosome movement of one pair of homologues is independent of any other pair of chromosomes. 39

information: the genetic specificity carried by DNA or messenger RNA as a linear sequence of nucleotides; the nucleotides are *aperiodic* (non-regular or non-repetitive in pattern) but not random. Each gene has its own unique sequence. 192

leaky mutant: a mutation which has lost most, but not all, of its function. 95

life cycle: the successive and cyclical stages of fertilization, development, maturity, and reproduction result in new generations with the eventual death or infertility of the older generation. 9, 16

lysogeny: a process in which bacterial cells harbor viruses in a quiescent state but occasionally disintegrate releasing detectable virus particles. 139

meiosis: a sexual process in specialized cells which reduces the chromosome number from diploid to haploid and, at the same time, scrambles the paternal and maternal genetic material within the cell. 5, 7, 13

metabolism: the sum of synthetic and degradative activities in a cell; the total activity of enzymes. 1, 8, 10, 12, 28, 73, 75, 153

microsome: a fragment of endoplasmic reticulum rich in ribosomes. 35, 185, 186, 193, 196

mitochondrion: a membranous structure in cell cytoplasm which contains the enzymes for oxidative reactions (*aerobic metabolism*) and provides the energy for multicellular forms of life. 2, 11, 35, 185

mitosis: the process by which one cell forms two cells of identical genotype. 2, 3, 7, 12

molecular biology: the study of biological macromolecules and their functions in cells and cell free systems. 5, 18, 85, 94, 149, 150, 174, 231

molecular disease: a concept that inherited diseases manifest symptoms which can be attributed to the properties of a defective enzyme or structural protein. 153, 167

morphogenetic movement: the migration of cells or tissues in the developing embryo. 101, 128, 129, 132, 134

mosaic development: an embryonic organization in which separated blastomeres or experimentally segmented portions give rise to incomplete or abnormal embryos. 99, 121

multiple factors: a theory that attributes continuously varying traits to the participation of several independent genes; also called polygenic inheritance. 268

mutation: a change in a gene by a physical or chemical means, resulting in the altered function of the gene product. 38, 64, 65, 68, 71, 73, 75, 80, 84, 88, 93, 128, 141, 143, 168, 171, 205, 206, 209, 213, 230, 239, 250, 283, 285, 291, 296

mutation theory: a view of evolution held by De Vries; it assumed that new species arose suddenly by discontinuous (non-gradual) mutations. 230, 238, 243

natural selection: a theory of evolution proposed by Darwin which asserts that naturally occurring variations exist in a population, and that more progeny are born than survive. The survivors are endowed with variations that make them more suited for life in their environment and gradually new species arise from this process. 70, 232, 235, 237, 243, 265, 305, 308

neo-Darwinism: in contemporary genetic thought, a theory of evolution embodying natural selection, survival of the fittest, and the origin of variation from gene mutations. 230, 268, 302

non-disjunction: the failure of homologous chromosomes to separate from one another in cell division. 66

nucleic acid: the chemical basis of heredity consisting of DNA or RNA. In most organisms, DNA carries the genetic information; it is transcribed into messenger RNA and decoded as protein. The subunits of nucleic acid are *nucleotides.* 27

nucleotide: the smallest functional unit of a nucleic acid; it consists of inorganic phosphate, a five carbon sugar, and a nitrogenous base. 1, 4, 93, 95, 168, 222

ontogeny: development or successive complexity from fertilization to birth. 31

operator: a region of an operon which initiates the reading or transcription of structural genes and enzymatic genes. 141

operon: a theory of genetic regulation which consists of regulatory genes and structural (or enzymatic) genes. The regulatory genes turn the other genes under its control to on or off states by feedback circuits involving biochemical repressors. 99, 138, 143, 151

organizer: A region of an embryo which can induce the formation of several tissues, including organ formation. 99

pangene: a 19th century unit of heredity, inferred on philosophic grounds; replaced today by the term *gene.* 29, 238

phenotype: the appearance of an organism; the sum of its genetic and environmental interactions. 6, 268

phylogeny: the relatedness of living things by a study of comparative anatomy, biochemistry, and genetics. 31

plasmagene: a discredited concept of cytoplasmic genes believed to cause development; now replaced by the operon model. 138

pleiotopism: the multiple effects of one gene, or the series of effects of a closely linked set of genes. 140, 141

preformation: the view that growth (enlargement) alone, not epigenesis, is the basis of development. The sperm, egg, or zygote would be a miniaturized form of the adult. 29, 30, 98

protein: a polypeptide composed of amino acids. *Structural* proteins form membranes and other components of a cell; *enzymatic* proteins carry out metabolism; and *regulatory* proteins serve to switch on or off the other types of genes. 2, 4, 175, 176, 180, 184

protein taxonomy: the view that classification of organisms can be achieved by analysis of the amino acid sequences of several strategic proteins common to all organisms. 180

pseudo-allelism: a concept of multiple alleles within which recombination takes place, separating segments which may be genes or portions of genes. 95

purine: a nitrogenous ring compound found in nucleic acids; in most organisms, guanine and adenine. 84, 85, 213

pyrimidine: a nitrogenous ring compound found in nucleic acids; in most organisms, cytosine, thymine, and uracil. 84, 85, 213

recombination: crossing over or the exchange of paternal and maternal segment of homologous chromosomes during meiosis. 9, 64, 94, 206

regulative development: a theory of embryology which attributes identical potentials to the blastomeres of an organism. Separated blastomeres form complete identical twins of one another. 99, 119

replication: the process of self-duplication of genetic material, especially by means of the Watson–Crick model of DNA. 84, 213

repulsion: the apparent separation of genes maintained by their presence in separate but homologous chromosomes; crossing over brings them into a coupled state. 62

ribonucleic acid (RNA): any of three forms of nucleic acid associated with genetic decoding and protein synthesis; *messenger* RNA carries the copy of genetic information from DNA; *transfer* or *soluble* RNA are synonyms for the substance which carries an amino acid to its proper codon; *ribosomal* RNA provides a rigid structure necessary for the protein synthesis. 3, 4, 8, 16, 33, 149, 150, 151, 183, 185, 189, 192, 193, 194, 197, 222, 223, 227

ribosome: a component of the endoplasmic reticulum rich in RNA; protein synthesis takes place *in vivo* and *in vitro* if ribosomes are present. 3, 4, 11, 33, 150.

segregation: the separation of genetic factors, without contamination, as a consequence of meiosis. 37, 39, 61, 280

soma: cells bearing a diploid number, which do not form gametes, and which are usually differentiated into tissues. 36

special creation: the view that all life as it exists now was present in the past and placed on earth in one short period of time by a Divine Being; specifically, special creation rejects the evolution of species and accepts the fixity of species. 229

syncytial mass: a segment of tissue containing many nuclei but lacking separate cell membranes. 102, 103

transition: a mutation caused by the replacement of one purine by another purine or of one pyrimidine by another pyrimidine. 205, 214

tryptic digest: the use of the enzyme trypsin to cleave proteins at the site of the amino acids arginine or lysine. Each such fragment forms a separate fingerprint on a chromatogram. 167, 168, 319

unit character: one of the fore-runner terms for *gene;* no longer used. 61, 262

uridine: the nucleotide found in RNA which contains the nitrogenous base uracil. Its pairing properties are similar to thymidine in DNA. 34, 35